Materials Selector For Hazardous Chemicals

MS-4: Hydrogen Flouride and Hydrofluoric Acid

Materials Selector Series for Hazardous Chemicals

Volume 1: Sulfuric Acid and Oleum
Volume 2: Formic Acetic and Other Organic Acids
Volume 3: Hydrochloric Acid, Hydrogen Chloride and Chlorine
Volume 4: Hydrogen Fluoride and Hydrofluoric Acid
Volume 5: Nitric Acid*
Volume 6: Ammonia and Caustic*
Volume 7: Phosphoric Acid*

*To be published

Materials Selector For Hazardous Chemicals

Original Transcript: *Tom F. Degnan*
Technical Editor: *Hira S. Ahluwalia Ph.D.*

MS-4: Hydrogen Fluoride and Hydrofluoric Acid

MTI Publication MS-4

Materials Technology Institute
of the Chemical Process Industries, Inc.

ISBN 1-57698-023-5

This document was prepared under the sponsorship of the Materials Technology Institute of the Chemical Process Industries, Inc. (MTI) and is approved for release. All data and information contained in this document are believed to be reliable; however, no warranty of any kind, express or implied, is made with respect to the data, analyses, or author of this document; and the use of any part of this document is at the user's sole risk. MTI, the author, or any person acting on its behalf, assume no liability and expressly disclaim liability, including without limitation liability for negligence, resulting from the use or publication of the information contained in this document or warrant that such use or publication will be free from privately owned rights.

Published by
Materials Technology Institute
www.mti-link.org

Contents

Foreword

The original basis for this monograph on materials selection and corrosion in the manufacture, handling and utilization of hydrogen fluoride and hydrofluoric acid was the NACE/MTI Materials Selection Advisor (MSA) prepared by Tom Degnan. A first draft was prepared and edited by C.P. Dillon. It has now been completely reviewed, updated and re-written by Hira Ahluwalia.

Information in this monograph is provided on the properties of hydrogen fluoride and hydrofluoric acid, production methods, health and safety issues, forms of corrosion, definitions, and relevant specifications for materials of construction, as well as pertinent laboratory and field corrosion data. Alloys are identified by their UNS number together with generic or trade name where appropriate.

The materials of construction preferred for manufacturing, storing, transporting, and handling hydrogen fluoride and hydrofluoric acid as a commercial product are described in detail.

Acknowledgment

MTI gratefully acknowledges the support of the Nickel Development Institute (NiDi) in producing the Materials Selector for Hazardous Chemicals (MS) series of books.

NiDi is an international nonprofit organization serving the needs of people interested in the application of nickel and nickel containing materials.

Head Office:
Nickel Development Institute
55 University Ave. Suite 1801
Toronto, ON M5J 2H7
www.nidi.org

1

Hydrogen Fluoride and Hydrofluoric Acid Properties and Hazards

Physical and Physiological Properties

Hydrogen fluoride and hydrofluoric acid are available in four grades: anhydrous liquid (AHF), 70% hydrofluoric acid, 48% reagent-grade hydrofluoric acid and 49% electronic-grade hydrofluoric acid. Some of the principal physical and physiological properties of hydrogen fluoride and of 70% hydrofluoric acid are as follows:

Physical Properties of Anhydrous Hydrogen Fluoride

Boiling Point	19.5°C (67.1°F)
Freezing Point	−83.5°C (−118.3°F)
Liquid Density @ 0°C (g/ml)	1.015
Gas Density (air = 1.00) @ 0°C	0.7
Conductivity, @ 0°C (mhos/cm)	1.6×10^{-6}

NIOSH* Permissible Exposure Limits

Time-weighted 8-hr average, ppm	3 (v/v)
Short time exposure limit, ppm	6 (v/v)
Immediately dangerous to life and health, ppm	20 (v/v)

Physical Properties of 70% Hydrogen Fluoride (Hydrofluoric Acid)

Boiling Point (760 mm/Hg)	66.6°C (151.9°F)
Freezing Point	−70.2°C (−94.4°F)
Density @ 25°C (77°F)	1.12 g/ml 10.18 lb/gal

Safety and First Aid

Hydrogen fluoride is poisonous to humans when inhaled and a corrosive irritant to skin, eyes, and mucous membranes. It produces severe, slow-healing burns. Concentrations of 50–250 ppm are dangerous for even brief exposures. Fluoride salts are poisonous, both *per se* and as corrosion products.

Hydrogen Fluoride must not be allowed to get in the eyes, on the skin, or on clothing. Do not breathe vapor or mist. Containers must be kept tightly closed. Wash thoroughly after handling. Personal safety and first aid is covered in detail in the Du Pont Material Safety Data Sheet H-13781–2 (Appendix A). For direct information, call Product Information (800) 441-9442. In a medical emergency, call (800)441-3637.

Speedy medical treatment is essential. The following first aid facilities and supplies should be available:

- Safety showers
- "Impervious" gloves
- Nebulizer
- 1% calcium gluconate solution
- Eye wash fountain—5-minute wash, followed by rinsing with 1% calcium gluconate solution as a continuous drip
- 2.5% calcium gluconate gel, applied by rubbing continuously while wearing an impervious glove
- "ZEPHIRAN" solution
- Oxygen (against potential lung damage)
- Milk of magnesia (in the event of ingestion)

First aid treatment must be initiated immediately by trained personnel, followed by prompt medical treatment.

Self-contained breathing apparatus (with a full face plate of a type approved by the Mine Safety and Health Administration or by NIOSH) should be readily accessible in areas in which leaks or spills may occur, and should be *worn* in areas where concentrations can exceed the permissible limits described above. For maximum protection against large leaks or in emergency situations, a fully encapsulating acid-resistant suit (one piece construction) with an independent air supply should be used.

Additional protective equipment that should be worn in an area where quantities of HF are present is as follows:

- Coveralls with long sleeves
- Chemical splash goggles or full-face shield
- Hardhat
- Gauntlets of polyvinyl chloride or neoprene
- Safety shoes with PVC, neoprene, or composition soles (Leather soles and gloves are not acceptable, since they can absorb HF.)

Where skin contact is possible, neoprene boots and neoprene or vinyl jacket and pants should be worn. Protective equipment should not be carried outside an operating area, but should be washed and removed in a prescribed order in accordance with a standard written procedure to avoid contact with any part of the body. In case of leaks, the self-contained breathing apparatus should be used.

Plans should be made for emergency escape from unexpected fume releases. Both eyes and lungs must be protected.

2

Hydrogen Fluoride and Hydrofluoric Acid Production

The principal uses of hydrogen fluoride and hydrofluoric acid are in the manufacture of chlorofluorohydrocarbon and fluorocarbon refrigerant fluids, aerosol propellants, specialty solvents, and high performance plastics. They are also used as a catalyst in alkylation of petroleum fractions to produce high-octane fuels and as an acidizing agent during injection of acid down oil and gas wells. Hydrofluoric acid is used in the separation and purification of uranium isotopes and in the manufacture of aluminum. It is an effective etchant for computer chips (electronic grade). Mixed with dilute nitric acid in small amounts (e.g., 1–3%), it is used as a pickling agent for steel, stainless steels, tantalum, and other metals.

Note that the electronic grade must be exceptionally pure and free of metal or organic impurities, including particulate matter. The total metal content must be less than 100 ppb, with no individual metal greater than 10 ppb. To achieve this end, vessels, piping, and ancillary equipment must give off hardly any metallic and/or organic substances or particulate matter into the media being transported (such as deionized water, acids, etc.). To this purpose, new techniques of fabrication as well as very high-purity polyvinylidene (PVDF) and perfluoralkoxy (PFA) components are required.[1]

Principal Manufacturing Processes

There are a number of manufacturing processes for the production of HF, based on different raw materials.

Fluorspar Processes

By far the greatest amount of HF is produced by the reaction of upgraded ("acid grade") fluorspar with concentrated sulfuric acid. There are a number of proprietary processes that use this reaction. In many cases, the details of operation have not been published, being held confidential. However, there is one process, licensed from Buss Ltd. of Basel Switzerland, that has been used by many companies worldwide, for which details have been published.

The Buss Process

In the Buss process, preheated, concentrated sulfuric acid and dried, ground acid-grade fluorspar are mixed and pre-reacted in a steam-jacketed interrupted-screw conveyer, called a Ko-Kneader(TM) Some HF is generated at this point and the reaction is completed as the materials pass through an externally heated rotary kiln. The gaseous HF generated in the kiln and the HF liberated in the pre-reactor mix and are fed through a scrubbing column where entrained solids are removed by a counter flow of condensate from an overhead condenser. This condensate of sulfuric acid, water, and other high-boiling chemicals are then mixed with fresh sulfuric acid and oleum to provide the acid feed to the pre-reactor. The crude HF vapor is condensed and is then refined in two columns, which removes impurities (such as SO_2, CO_2, and H_2SO_4).

The Ko-Kneader is made of Alloys B-2 and C-276 or C-4. The kiln and parts of the distillation column are made of steel. The scrubbing section of the pre-condenser, the bottom section of the second distillation column, the sulfuric acid preheater, and the intermediate tank, where corrosive mixtures of HF, H_2SO_4, and H_2O are handled at temperatures of 90°C (194°F) or higher, are made of high-nickel alloys.

Other Fluorspar Processes

Other processes that make HF by the reaction of fluorspar with sulfuric acid are as follows:

- Rotary Kiln
- Agitated Retort
- Ball Mill
- Conveyer Mixer Reactor
- Continuous Reactor (heated through agitator)

Kiln type reactors, except for the Buss process, generally require high-nickel alloy liners (e.g., Alloy C-276, Alloy G-3, or Alloy 625 at the feed end where sulfuric acid is introduced and at the discharge end, where sulfuric acid condenses.

HF from Fluosilicic Acid

Fluosilicic Acid (FSA; $H_2[SiF_6]$) is a byproduct from the "Wet Process" manufacture of phosphoric acid, phosphates, and phosphate fertilizers in which phosphate rock is treated with sulfuric acid. FSA is considered a useless byproduct and is difficult to dispose of. Because phosphate rock contains 3–5% fluorine, significant quantities of FSA are available.

Two processes are the U.S Bureau of Mines process and the Buss process.[2]

In the Bureau of Mines process, lime, silica, and FSA are mixed and the resulting calcium fluoride and silica precipitated, filtered, and pelletized. The pellets are fed to a 565°C (1050°F) kiln. The HF-water vapor stream is condensed and saturated with sodium fluoride. The sodium bifluoride salt is filtered, dried, and heated to 400°C (750°F). The salt dissociates to HF and sodium fluoride. The HF is condensed and collected.

In the Buss process, the FSA is concentrated, vaporized, and enters the middle of an absorption column in the form of an HF-SiF_4-H_2O gas mixture. In the upper half of the column, an organic solvent selectively absorbs HF. The HF is recovered in a vacuum evaporator.

Two other processes used are the Harfluor and Wellman-Power Gas processes. Although a number of processes have been developed, finding a phosphate processor with sufficient FSA to support an HF plant has been a stumbling block to commercial production.

References

1. E. Werthmuller, Paper No. 341, CORROSION 90, Houston, TX (NACE International, Houston, TX).
2. W.R. Rogers and K. Muller, Chemical Engineering Progress, Vol. 59, No. 5, May 1963, p. 85–88.

3

Corrosion in Anhydrous Hydrogen Fluoride

Anhydrous hydrogen fluoride (AHF), defined as <400-ppm water by Hydrogen Fluoride Industry Practices Institute (HFIPI), is a clear, colorless, corrosive liquid with a sharp, penetrating odor, with an atmospheric boiling point of 19.5°C (67.1°F). Aqueous HF is commercially available in bulk shipments at 49 and 70% concentrations, while smaller quantities may be available in other concentrations. HF will give a constant boiling mixture (azeotrope) at 37.7 weight % and 111.4°C (232°F).

The corrosion characteristics of various materials specifically in AHF are as follows.

Carbon and Low Alloy Steels

Liquid anhydrous hydrogen fluoride is commonly transported in carbon steel pipelines, stored in carbon steel storage tanks, and shipped in carbon steel tank cars. The corrosion rates of carbon steel in AHF are less than 10 mpy in the liquid phase and 5 mpy in the vapor phase at temperatures up to 66°C (150°F). A thin, protective iron oxide-fluoride film forms. Closely fitting steel parts may become cemented together. Laboratory data often show 10 times these rates of attack, reflecting the initial high corrosion rate while the protective film is still being developed.

Laboratory corrosion data indicate that the upper temperature limit for carbon steel in HF vapor is about 300°C (520°F). In addition, kilns contacting HF vapor at 300°C (520°F) have done so for many years without excessive corrosion.

The corrosion rate of steel in liquid and vapor AHF at various temperatures is shown in Figure 3.1. These data were generated in stagnant 100-hour (average) tests.

Erosion-Corrosion

Carbon steel has good erosion-corrosion resistance to liquid AHF at ambient temperatures. A design velocity limit of 1.8 mps (<6 fps) has been found to be conser-

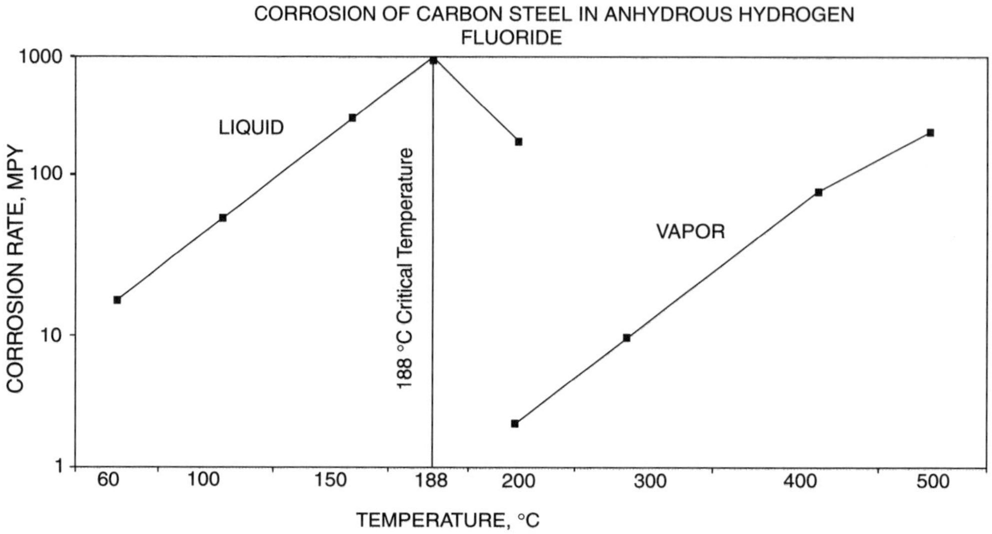

Figure 3.1 Corrosion of Carbon Steel in AHF Liquid and Vapor Curve over a Range of Temperatures

vative, although Comite Technique European Du Fluor (CTFE) suggests a limit of 1.5 mps (<5 fps).[1] These limits are suggested to minimize damage to the protective film by the flow.

Erosion-corrosion failures have been reported where direct impingement has occurred at tees. They have also been experienced during start-up if the equipment was washed out with lime water and started up without the opportunity for removal of all water or for the steel to form a passive film. (See section on Passivation below.)

Hydrogen Damage

Although carbon steel has good corrosion resistance, it is susceptible to several forms of hydrogen damage. To avoid serious problems, it is necessary to select the proper steel quality, fabrication procedures, design and inspection as described further in Chapter 7. When AHF is in contact with carbon steel, the following reactions occur:

$$Fe + 2\,HF \rightarrow FeF_2 + 2\,H^o \text{ (atomic hydrogen)}$$
$$2\,H \rightarrow H_2\uparrow$$

However, HF (and the small amounts of sulfides and arsenic impurities present in the steel) may suppress the second reaction, acting as so-called "poisons." As a result, the atomic hydrogen can enter the steel lattice, migrating to voids, dislocations, and inclusions, where it can dimerize to molecular hydrogen and cause blistering, delamination, hydrogen embrittlement (HE), hydrogen stress cracking (HSC), and several forms of hydrogen-induced cracking (HIC).

Hydrogen Blistering Hydrogen blistering has been observed and documented in AHF storage tanks.[2,3] Hydrogen blistering is most likely to occur in silicon-killed steels (such as SA 515) or more particularly with fine grain silicon-aluminum killed steels (such as SA 516 and SA 442), which are otherwise desirable because of low temperature toughness qualities. It occurs less frequently in semi-killed steels.

Hydrogen blistering and HIC can be minimized by using steels (particularly SA 516) with low sulfur (0.006% max.) and with inclusion shape control, as further described in Chapter 7. As a precaution against blistering caused by laminated areas, plates can be specified to meet ultrasonic testing requirements. Hydrogen blistering damage is not reversible.

Hydrogen Embrittlement (HE) HE is a loss of ductility in steel, under slow strain rates (i.e., it is undetectable by impact testing), due to dissolved atomic hydrogen. HE damage is reversible. Ductility can be restored by heating the steel to a temperature of 260–279°C (500–700°F) to permit the H^o to escape from the interstices of the metal lattice.

Hydrogen Stress Cracking (HSC) HSC occurs in welds and HAZs when the local hardness exceeds about HRC 22. Like blistering, HSC is not reversible because of the physical damage. However, it can be prevented by proper selection of steels and welding parameters or by proper post-weld heat treatment (PWHT) to reduce residual stress and hardness.

Hydrogen-Induced Cracking (HIC) HIC, including stress-oriented hydrogen-induced cracking (SOHIC) which is also known as "stepwise cracking" (SWC), can originate as HSC in hardened welds or HAZs or at discontinuities and inclusions in the steel. HIC can proceed in a stepwise manner as cracks in different planes link up. The resulting SOHIC can form through-wall cracks. This is prevented by specifying clean steels, as described in Chapter 7. HIC and SOHIC also constitute permanent physical damage and are not reversible.

Effects of Alloying Elements and Heat Treatment

Alloy steels have been generally reported to have higher corrosion rates than carbon steel, in aqueous and anhydrous HF at temperatures *above* ambient. The alloying elements produce a loose, porous, non-protective corrosion product layer and cause increased corrosion in AHF.

A fluorochemical plant had corrosion of about 25.4 mm/y (1000 mpy) on pipe, while carbon steel fittings on either side had only 0.89–1.27 mm/y (35–50 mpy). The conditions were about 10% AHF and 90% organics at 80°C (176°F). The failure investigation showed the pipe to be fabricated from a low temperature alloy steel per

ASTM A333, grade 9 (UNS K22035) with copper (0.81%) and nickel (1.86%). The fittings had no copper, nickel, or chromium.[4]

Table 3.1 shows data from in-plant and lab AHF corrosion studies of steels with varied alloy compositions.[4] This study indicates that steels with copper, nickel, and chromium are susceptible to higher corrosion in AHF and the corrosion rate increases with temperature. The researchers also observed higher corrosion rates in areas of turbulence.

At temperatures higher than 66°C (150°F), where corrosion of carbon steel in liquid AHF becomes appreciable, normalized carbon steel forgings may be more corrosion-resistant than annealed. This allegation is based on a verbal report of a premature failure of weld neck flanges in an HF alkylation plant piping system. The failed flanges were found to have an annealed microstructure. Flanges are usually normalized, and normalized flanges are sometimes required by company specifications. Normalized flanges did not show much corrosion. No alloying elements were found. As in sulfuric acid, accelerated corrosion was attributed to the presence of a spherodized carbide structure, which would have been eliminated by normalizing. Note that ASTM A-105 flanges 4″ ips and under, may not be heat treated since hard microphases may lead to HE-type cracking.

Passivation

Although the corrosion rate of carbon steel in AHF below 66°C (150°F) is low after the thin protective film forms, the initial corrosion rate is much higher. This film may not form at all under conditions of excessive velocity. To prevent erosion-corrosion failures, piping and equipment should be filled initially with AHF for some period of time (a day should usually be sufficient) to allow the film to form. This should be done both initially and before equipment or pipelines that have been washed out are placed back in service.

Table 3.1 Carbon and Alloy Steels Exposed to AHF Liquid, 40–90°C Corrosion Rates: mm/y (mpy)

ASTM Designation	Minor Element Composition (wt%)			2 Week Lab Test 100% AHF 90°C	1.3Y, Plant Test 100% AHF 75–80°C	1.3Y, Plant Test 100% AHF 40–45°C
	Cu	Ni	Cr			
A 516 Grade 55	0.03	0.02	0.02	1.19 (47)	0.55 (22)	0.12 (4.7)
A 526 Grade 55	0.09	0.07	0.06	1.88 (74)	—	—
A 106 Grade B	0.28	0.11	0.10	2.79 (110)	0.77 (30)	0.11 (4.4)
A 588 Grade A	0.35	0.61	0.20	—	>0.94 (>37)	0.14 (5.4)
A 203 Grade E	—	3.47	—	4.09 (161)	>1.14 (.45)	0.17 (6.6)
A 710 Grade A	1.05	0.7	0.68	5.54 (218)	>1.27 (>50)	0.19 (7.3)

It should be noted that DOT regulations for shipping AHF and 70% HF *require* that tank cars or motor vehicles be "passivated" with the acid being shipped or by an equally effective method before being placed in service and after being washed with water.[5]

Austenitic Stainless Steels

Austenitic stainless steels, including Type 304, Type 316, and their cast equivalents (CF-8:J92500 and CF-8M:J92600, respectively) have good corrosion resistance to liquid AHF at up to 100°C (212°F). Some unpublished Du Pont data show corrosion rates of less than 1 mpy (0.13 mm/y) for Type 304 at 100°C (212°F), but 29 mpy (0.76 mm/y) at 150°C (300°F). Another investigator ran a series of corrosion tests on austenitic stainless steels, including Types 302, 317, and cast Alloy 20 at temperatures from 27°–88°C (80°–190°F) with corrosion rates averaging between 2.4 and 6.2 mpy (0.06–0.16 mm/y).[2] Still others reported a corrosion rate of 2 mpy (0.15 mm/y) for Type 316 at 28°C (100°F) in 98% HF.[6]

Type 316 is commonly used for instruments such as manometers. CF-8M (J92600) is often used for valves and pumps at ambient or lower temperatures. Both Types 304 and 316 have been used for bolting because of their superior atmospheric corrosion resistance, but Types 301, 302, and 303 should be avoided, as described further in Chapter 7.

Types 304 and 316, like carbon steel, have been known to fail by direct impingement or severe cavitation. In erosion-corrosion tests at 5°C (41°F) at high velocities (13–24 fps), the corrosion rates were 12–14 mpy, about half that for carbon steel, but substantially higher than for nickel alloys.

An MTI manual indicates that austenitic stainless steels resist stress corrosion cracking by fluorides.[7] However, other investigators have shown that the threshold chloride ion concentration necessary to cause intergranular stress corrosion cracking (IGSCC) of *sensitized* Type 304 is considerably reduced by the presence of fluorides.

High nickel stainless steels containing copper, including wrought Alloy 20Cb-3[7] and Alloy 825, have excellent corrosion and erosion-corrosion resistance to liquid AHF, superior to Types 304 and 316. These alloys are particularly useful in handling HF-60% H_2SO_4, as described in Chapter 6.

Cast Alloy 20 is an excellent choice for valves, valve trim, and pumps. The alloy has been somewhat difficult to cast but castability has been improved in recent years. There is a documented report of severe intergranular cracking of cast stub ends exposed to a 96% HF stream containing BF_3 at 0°C (32°F). It is now reported that CN7M cast from AOD metal has better corrosion resistance and castability.[8]

Also, IGSCC of Type 316 equipment, initiating in the HAZ of welds, has been reported in BF_3 alkylation. The vessel was replaced in Type 316L and was thermally stress-relieved at 870–900°C (1600–1650°F) for 4 hours.

Nickel-Based Alloys

There is an inherent difference among nickel-based alloys (i.e., those containing more than 50% nickel) depending upon whether or not they contain chromium. (Cr-free alloys tend to resist reducing acids while the Cr-bearing grades resist oxidizing conditions.)

Chromium-Free

Cr-free nickel alloys have excellent corrosion and erosion-corrosion resistance to AHF, both liquid and vapor. Data for corrosion tests under velocity conditions in liquid AHF for a number of nickel alloys are shown in Table 3.2. However, these alloys (more particularly weld deposits that contain niobium or titanium) are attacked at appreciably higher rates at elevated temperatures or in aqueous HF as described in Chapter 5.

Alloy 400 (Monel 400; N04400) Alloy 400 and its cast version N24025 are the least expensive and most popular nickel alloy used in AHF service. They are commonly used for valves, valve trim, and pumps. Alloy 400 is also used for process piping and equipment at temperatures above the useful range for carbon steel (certainly above 100°C [212°F]) for AHF.

If Alloy 400 equipment is exposed to moist air in the vapor phase, even occasionally, aqueous HF formed may cause SCC. The responsible agent is cupric fluoride (CuF_2). Selection of alloy resistant to SCC, such as Alloy 600, may be required, as described further below. Alternatively, the equipment may be thermally stress re-

Table 3.2 Erosion-Corrosion in AHF; 50-Hr Tests 5°C

Materials	0 fps	10 fps	20 fps
Steel (K01700)	7	19	28*
Type 304	1	10	13*
Type 316	1	16	15*
Alloy 20Cb37	—	6	4**
Alloy 825	—	6	2
Alloy 200	—	9	13**
Alloy 400	—	3	7**
Alloy 600	—	7	7*
Alloy B	—	2	3**
Alloy C	—	1	3

*Average of 3 tests
**Average of 2 tests

lieved, although this may not be entirely effective, as discussed further below for aqueous systems.

Vaporizers, normally tube-and-shell heat exchangers, have been constructed of both Alloy 400 and Alloy 600. Vaporizers can contain air, making Alloy 600 a better choice. However, vaporizers of both metals have failed in crevices, under deposits, and at liquid-vapor interfaces.

Corrosion has been attributed to concentration of residual sulfuric acid, which is normally present at a concentration of about 10 ppm (50 ppm max. by specification), and to sodium hydrogen fluoride (NaHF). Deposits have been found to be largely ammonium arseno-fluoride (NH_4AsF_6), although the concentration of NH_3 in AHF is usually less than 1 ppm.

The primary solution to the problem has been to reduce the temperature of the heating medium. In one case, 50 psig steam was used in place of 150 psig steam.

Nickel 200 is also satisfactorily resistant to AHF, although the corrosion rates are slightly higher than for Alloy 400. Some comparative corrosion data from the literature shows rates of 1–3 mpy for Alloy 400 at 16–38°C (61–100°F) vs. 3 mpy for Alloy 200 at 16–27°C (61–81°F).[9] At slightly higher temperatures of 33–50°C (91–122°F) for Alloy 400 and 70°C (158°F) for Alloy 200, both materials showed rates of 5 mpy. Some other comparative data are shown in Table 3.3.[10]

Alloy B-2 is resistant to AHF but there is no occasion to use it in preference to the lower nickel alloys.

Chromium-Bearing

The Ni-Cr-Fe Alloy 600 has sometimes been used in lieu of Alloy 400 for vaporizers, which are normally horizontal shell-and-tube heat exchangers. Vaporizers can experience ingress of air, which makes Alloy 600 a better choice, although vaporizers of both metals have failed in crevices, under deposits, or at liquid-vapor interfaces.

The high-performance alloys in this category comprise the wrought Ni-Cr-Fe-Mo-Cu and Ni-Cr-Mo groups. The first group includes Alloys G-3 and G-30; the second, Alloys 625, C-276, C-22 , C-2000, 686, and 59.

All of these alloys have excellent corrosion and erosion-corrosion resistance to AHF, liquid or vapor. However, they offer little if any advantage in these environ-

Table 3.3 Corrosion Rate of Alloy 400 and Alloy 200 in AHF at Various Temperatures

Material	Temperature	Corrosion (mpy)
Alloy 400	50°C (122°F)	1*
Alloy 400	100°C (212°F)	1*
Alloy 400	122°C (250°F)	0.5
Alloy 400	300°C (570°F)	1
Alloy 200	300°C (570°F)	3

*Agitated by stirring @ 1750 rpm

ments over Alloy 600 and Alloy 400, except under conditions of abrasion and high temperature. A number of these alloys, however, have superior resistance to the hot corrosive mixtures encountered in HF manufacture (see Chapter 5).

Alloy C-276 is the most readily available of these alloys in bellows and instrument parts and has been used in place of Alloy 400, the latter being susceptible to SCC if exposed to moisture and air without being thermally stress relieved.

Hot-wall corrosion tests conducted simulating AHF vaporizer conditions were conducted, with the metal-wall temperature at 140°C (284°F), in boiling AHF at 70°C (158°F). The results showed that, although the prototype Alloy C, Alloy C-276, and Alloy 625 specimens suspended in the boiling AHF all had corrosion rates of less than 1 mpy, the Alloy C heated disc had greater corrosion attack than did the Alloy 600 and 400 discs both at the liquid-vapor interface and also under the gasket.

Copper and Copper Alloys

Although laboratory corrosion tests under static conditions indicate that pure copper has acceptable corrosion resistance to both liquid and gaseous AHF to 200°C (390°F), field tests and experience in HF alkylation have demonstrated that copper and its alloys, except for the copper-nickels, are velocity sensitive.[9] They are corroded rapidly at velocities associated with bubble caps in distillation units and with heat exchangers. Yellow brass has been reported to be subject to stress corrosion cracking (SCC).

In an HF alkylation column, operating at 66°C (150°F), bubble caps of silicon bronze (C65500) and copper corroded away at the rate of 120 mpy, while Alloy 400 caps were essentially unaffected.

The cupronickel alloys, particularly 70–30, are not velocity sensitive and are used for heat exchangers and equipment in AHF service.

Some data obtained in 100-hour static bomb test are presented in Table 3.4.

In other tests, metal coupons were exposed, suspended in test chambers for from 6 to 40 days. The results are summarized in Table 3.5.[11]

Table 3.4 Corrosion Rate of Copper-Nickel Alloys at Various Temperatures

	Corrosion Rate (mpy)		
Material	100°C (212°F)	150°C (300°F)	200°C (390°F)
Copper ()C11000	3.5	2.5	7.8
70–30 CuNi	1.4	8.5	1.9
Alloy 400	1.2	Nil	0.7

Table 3.5 Corrosion Rate of Various Copper Alloys

Material		Corrosion Rate (mpy)		
		16–27°C	27–38°C	82–88°C
Copper	C11000	13	—	—
70–30 Cupronickel	C71500	—	2.0	10
Alloy 400	N04400	3.2	0.9	—
Red Brass	C23000	30	16	50
Admiralty B Brass	C44300	10	13	20
Aluminum Bronze	C61300	14	—	—
Phosphor Bronze	C51000	20	19	60

Precious Metals

Despite their high initial cost, several of the precious metals have found practical application in AHF.

Silver Silver, including fine (P07020) and sterling (P07931) grades, has a high degree of corrosion resistance to AHF liquid and vapor. It has been widely used for laboratory equipment to withstand AHF. An 88-hour test at 38°C (100°F) in AHF liquid resulted in a corrosion rate of 1.8 mpy.[12]

Silver clad steel HF regenerators were used in the first UOP HF alkylation plants, based on laboratory corrosion tests.[13] Not until the plant was in operation was it discovered that AHF exposed to hydrocarbons containing only traces of sulfur would absorb all of the sulfur. The sulfur in the acid then attacked the silver, converting it to silver sulfide.

It has also been reported that low silver (50%) solders (i.e., with UNS numbers below P07490) were attacked rapidly by flowing AHF.[9] For exposure to flowing AHF in laboratory equipment, ASTM grade No. 7 (70 Ag- 20 Cu-10 Sn) silver brazing alloy has been successfully used.

Other Precious Metals Gold, platinum, and their alloys are very resistant to AHF, as well as to aqueous HF. Platinum rupture discs have been used in HF manufacture and show no corrosion.

Members of the platinum group of metals, including iridium, palladium, and rhodium are also resistant to AHF. Palladium may be subject to attack in aerated aqueous acid and in mixtures of HF and concentrated sulfuric acid, based on its behavior in aggressive acidic environments.

Other Metals

Molybdenum and *tungsten* both have a high degree of resistance to both anhydrous and aqueous HF. Molybdenum valve trim has been used to combat erosion-corrosion in handling hot crude AHF. Tungsten had a corrosion rate of 0.1 mpy in 70°C (158°F) AHF. Unfortunately, both metals are relatively brittle and difficult to weld.

Aluminum and *magnesium* have shown corrosion rates of less than 10 mpy at 100°C (212°F) in one series of tests (see Chapter 4). The Aluminum Association's publication, "Aluminum with Food and Chemicals," claims that "Aluminum is resistant to hydrogen fluoride in the gaseous and liquid states. However, both metals have very high corrosion rates with lower strength HF, which precludes washing the equipment with water."

One investigator conducted a series of laboratory corrosion tests in 80–99.5% HF at room temperature on unalloyed aluminum (A91100), aluminum alloyed with manganese (A93003), and with magnesium (A96063).[14] Unalloyed aluminum had corrosion rates of 2 mpy or less and the manganese alloy of less than 5 mpy in these concentrations. The magnesium alloy had corrosion rates exceeding 50 mpy at concentrations of 80–85%.

Aluminum and magnesium should be used with the greatest caution. Field corrosion tests in AHF showed erratic results for both metals.[10] Aluminum had a corrosion rate of 20 mpy in the temperature range 16–20°C (61–68°F) and 976 mpy at 82–88°C (180–190°F). Magnesium (Dow metals E and J) had corrosion rates of 5–20 mpy at 16–27°C (61–81°F) and nil at 54°C (129°F). Their behavior in AHF parallels their performance in hydrogen chloride, except that the corrosion rate in aqueous hydrofluoric acid is greater for aluminum than for equal concentrations of hydrochloric acid.

Other metals reportedly resistant to AHF include thorium and mercury.[15]

Nonmetals

Nonmetallic materials of construction are used for gaskets, packing, valve diaphragms, plug cock sleeves, pump seals, and hose lining for tank car unloading, as well as for vessels, columns, pumps, and linings for piping and equipment. Because of the hazardous nature of HF, they should not be used for pressure vessels or pressurized piping unless contained in a ductile metallic structure.

Plastics The only plastics with satisfactory resistance to AHF are fluorinated thermoplastics. Temperature limitations depend upon the degree of fluorination.

Fluorocarbon plastics are widely used for gaskets, lined hose, valve parts, and other applications in AHF service. Composites formulated with glass or silica should be avoided.

The following fluorocarbon plastics have chemical resistance to AHF at temperatures up to their normal temperature limitations. However, creep, permeability, and differential thermal expansion must also be considered.

- PTFE Polytetrafluoroethlyene
- FEP Fluorinated Ethylene Propylene

- PFA Perfluoroalkoxy
- CTFE Chlorotrifluoroethylene

PVDF Polyvinylidene Fluoride is recommended by the manufacturer for AHF service to 93°C (200°F). There has been a failure reported in crude AHF containing concentrated sulfuric acid, for which service PVDF is not recommended. ETFE (Ethylene Tetrafluoroethylene) is recommended by the manufacturer to 110°C (230°F) in AHF.[16] ECTFE (Ethylene Chlorotrifluoroethylene) has been recommended by one source to a maximum temperaure of 115°C (240°F) in AHF although the manufacture has not conducted confirmatory tests.[17]

Elastomers Fluoroelastomers are available that resist AHF at room temperatures.[18,19] The optimum grade and curing agent must be selected for the service. There are many grades or series of polymers within each product. Avoid magnesia curing and silica, barium sulfate, titanium dioxide, and diatomaceous earth additives.

Perfluoroelastomers are resistant to at least 100°C (212°F), but are considerably more expensive.[20] As with fluoroelastomers, grade and curing agents are important. Kalrez(TM) grade 4079 is recommended for AHF.

These elastomers are used for 0-rings, diaphragms, tubing, and packing.

Ceramics High purity alumina (min. 99.7%) is resistant to both aqueous and anhydrous HF. Alumina has been used for pump seals running against carbon in liquid AHF service.

Nickel-bonded tungsten carbide-faced pump seals have been used successfully in AHF service. However, cobalt bonding is leached out by HF and should not be used.

Silica-free silicon carbide showed no attack in aqueous HF and in HF-nitric acid at ambient temperature and is presumed to be resistant to AHF.[21] It is used for pump components (seals, bearings, and sleeves), valve trim and heat exchanger tubes. This product is available as "Hexoloy" SA 1(TM) silica-free silicon carbide (registered trademark of Sohio Engineered Materials Co.).

Carbon and Graphite High-purity (low ash) carbon is suitable for pump seal parts and other non-pressurized applications. Flexible graphite has been recommended for gaskets.[1]

References

1. "Materials of Construction for Anhydrous Hydrogen Fluoride," Draft #4 (2/91). HF Panel of the Chemical Manufacturers Association.
2. F.A. Prange, Corrosion, Vol. 8, No. 10, p. 355 (1952).
3. R.L. Schuyler III, Materials Performance, Vol. 18, No. 8, p. 9 (1979).
4. T.F. Degnan. "Process Industry Corrosion—the Theory and Practice," National Association of Corrosion Engineers, p. 285 (1986).

5. "Hazardous Materials Regulations Pertaining to Hydrofluoric Acid and Hydrogen Fluoride," (49CFR), Parts 178–179. Department of Transportation (DOT).
6. W.Z. Friend and H.O. Teeple, Oil and Gas Journal, 16 March 1946.
7. D.R. McIntyre and C.P. Dillon, "Guidelines for Preventing Stress Corrosion Cracking in the Chemical Process Industries," MTI Manual No. 15, Materials Technology Institute of the Chemical Process Industries, Inc., St. Louis, MO (1985).
8. J.L. Gossett, "Specifications for Obtaining High-Quality, High-Nickel Castings C CW-2M and AOD-Refined CN-7M," Materials Performance, Vol. 28, No. 1 (January 1989).
9. Philips Petroleum Company, "Hydrofluoric Acid Alkylation" (1946).
10. International Nickel Co. Inc. "Corrosion Engineering Bulletin" CEB-5 1975.
11. M.E. Holmberg and F.A. Prange, Industrial Engineering Chemistry, Vol. 37, p. 103O (1945).
12. H.H. Uhlig, "Corrosion Handbook," John Wiley and Sons p. 316 (1948).
13. D.G. Bergman and G.W.G. McDona1d, Corrosion, Vol. 16, No. 1, p. 9 (January 1961).
14. V.S. Zotikov et al., Zashik Metallov, 10, 1974, p. 164.
15. E. Lingnau, Werkstoffe und Korrosion, No. 4, p. 216 (1957).
16. E.I. DuPont DeNemours & Co. Bulletin, "Hydrofluoric Acid Anhydrous—Technical Properties, Uses, Storage and Handling."
17. Fusibond Piping Systems, Inc. "Fusibond Plastic Lined Piping Systems—Chemical Resistance and Liner Selection Guide."
18. E.I. Du Pont De Nemours & Co. "Viton® Fluoroelastomer Fluid Resistance Guide" VT-500.1, 9/90.
19. 3M Co. "Fluorel® Fluoroelastomer Chemical Resistance" 4/86.
20. E.I. Du Pont De Nemours & Co. "Kalrez® Perfluorocarbon Chemical Resistance and Fluid Compatability," 10/1/88.
21. Carborundum Co. "Corrosion-Erosion Resistant Components for the Chemical Processing Industry," 5/85.

4

Corrosion in Aqueous Hydrogen Fluoride, Hydrofluoric Acid

Hydrogen fluoride is readily soluble in water over a wide range, the water solutions being referred to as hydrofluoric acid. Commercially, this acid is marketed in concentrations from 30-70%, the latter being a common grade. An azeotropic mixture of 38.2% occurs, having a boiling point of 112°C (234°F). Thus concentrations greater than 38.2% cannot be produced by evaporation and must be made by absorption of anhydrous hydrogen fluoride (AHF) in water.

It is interesting to note that laboratory characterization of nickel-based alloys in hydrofluoric acid has not been as extensive as it has with other common, inorganic acids. This is largely because tests using hydrofluoric acid cannot be run in standard laboratory equipment and because of the dangerous nature of hydrofluoric acid. While reviewing the data presented in this chapter, it is important to note that short-term weight loss tests in hydrofluoric acid are not as highly reproducible as in the case of other acids such as hydrochloric acid.

Aqueous solutions of hydrogen fluoride, hydrofluoric acid may be radically different from AHF in their corrosion behavior, even in the absence of contaminants or admixtures.

The corrosion characteristics of various materials specifically in aqueous HF are discussed in this chapter.

Metals and Alloys

Carbon and Low Alloy Steels

Carbon steel tanks and tank cars may be used to ship hydrofluoric acid of 65% strength or higher.[1] In practice, the only aqueous grade shipped in steel is 70%. Chlorobutyl elastomer or fluorocarbon sheet linings are used to prevent metallic contamination, as described further below.

All of the material, heat treatment, fabrication, and quality control requirements for AHF tanks and piplines are necessary (see Chapter 7). In addition, the following requirements are recommended:

- Maximum temperature 32°C (90°F).
- Minimum corrosion allowance ⅛″.
- Passivation (Chapter 3) before initial startup and after washing out with water.
- Maximum velocity of 2 fps (0.8 mps)
- Avoid dilution of heels by entry of moist air. Steel tank cars should be unloaded using nitrogen or dried air, for example.
- Avoid galvanic coupling, particularly with Alloy 400 or copper-nickel alloys.

Some corrosion test data for carbon steel in aqueous HF for 25 days at room temperature are tabulated in Table 4.1.[2]

Data from another source, presented in Figure 4.1, gives analogous data for tests conducted at 21–38°C (70–100°F). These latter tests were run at ambient temperature under conditions of low flow and absence of oxygen or oxidants.

However, the reported corrosion rates for 63–70% acid are too low. Other tests showed carbon steel to have a corrosion rate of 40 mpy in unaerated 70% HF at 52°C (125°F), based on 335 hour tests.[3] This corrosion rate corresponds closely to that experienced by steel tank cars used for shipping 70% HF. The latter tests also showed severe galvanic corrosion of steel coupons welded to more noble metals.

Alloy steels, particularly those containing chromium, suffer accelerated corrosion rates in dilute HF.[2] Comparative corrosion data with other mineral acids in 3-hr. tests at 45°C (113°F) are shown in Table 4.2.[2]

Actually, such rates are ridiculous from the practical standpoint, serving only to illustrate the diminution of corrosion resistance in reducing acids caused by chromium additions.

Austenitic Stainless Steels

Austenitic stainless steels have limited resistance to aqueous hydrofluoric acid. Type 304 has poor resistance to any significant concentration, but shows markedly less

Table 4.1 Corrosion Rate of Carbon Steel in Various Concentrations of Hydrofluoric Acid at Room Temperature

Concentration (%)	Corrosion Rate (mpy)
58	120
60	100
61	82
62	59
63	10
64	2
65	2
67.5	2
69.9	3

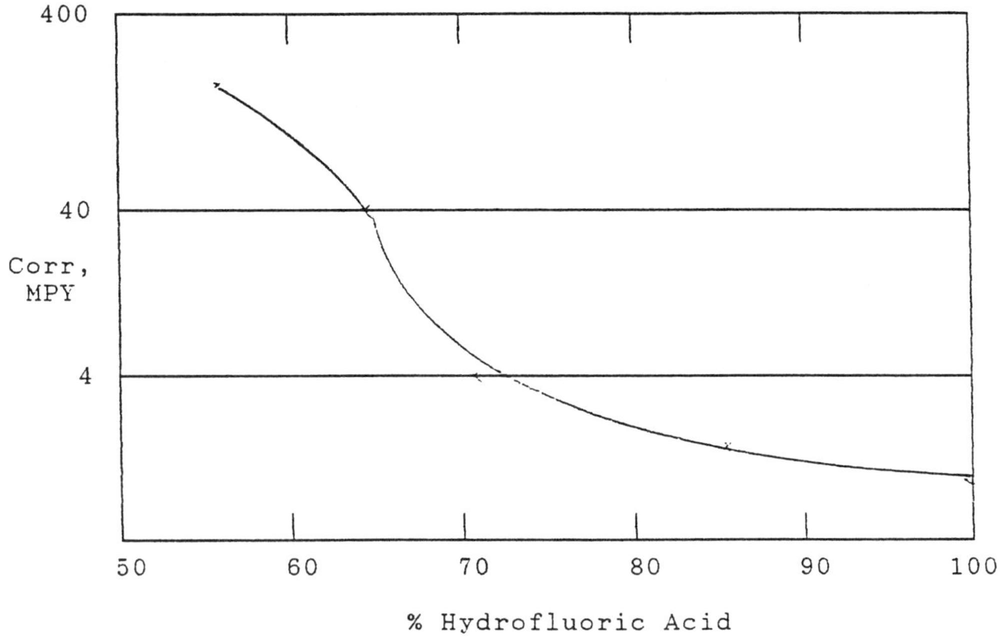

Figure 4.1 Corrosion Rate of Carbon Steel in Hydrofluoric Acid

Table 4.2 Corrosion Rate of Iron and Steel in Mineral Acids (mm/y)

Material		4.4%	7.4% H_2SO_4	5.1% HCL
Armco Iron		14	22	13
Carbon steel ASTM A106		26	20	23
ASTM A335	P11 (12Cr-3Mo)	290	41	4
	P12 (2 Cr-1 Mo)	640	330	10
	P5 (5 Cr-2 Mo)	290	94	3
	P9 (10 Cr-1 Mo)	330	53	6

corrosion than carbon steel external bolting in HF manufacturing plants. Type 316 has some resistance to dilute acid (<10%) at ambient temperature but is subject to pitting and is not recommended. These alloys have much poorer resistance than carbon steel in 70–90% acid concentration.

The high-performance stainless steels containing copper, e.g., Alloy 20 and Alloy 825, have limited resistance to all concentrations at ambient temperatures and up to 10% to 70°C (158°F). Cast Alloy 20 pumps and valves can be used for occasional service in 70% HF at ambient temperatures (e.g., for tank car or tank truck unloading), but corrosion rates are excessive for continuous service.

Some laboratory corrosion test data are shown in Table 4.3 (corrosion rates in mpy).[4]

Additional short term data[5] generated for a number of stainless steels, including the super-austenitic stainless steels, are presented in Table 4.4. The adverse effect of increasing the test temperature and concentration is clearly shown. The superaustenitic alloy Al6-XN shows useful resistance at 20°C up to a concentration of 5 molar HF.

For information on stress corrosion cracking (SCC) of stainless steels and hydrogen embrittlement (HE) of work-hardened stainless steels, see Chapter 7 under Bolting. In summary, austenitic stainless steels in the solution annealed condition are resistant to transgranular stress corrosion cracking. However, intergranular stress corrosion cracking (IGSCC) can occur in sensitized stainless steels in the presence of fluorides.

Table 4.3 Corrosion Rate of Various Austenitic Stainless Steels in Hydrofluoric Acid (mpy)

% HF	Temp°C	Days	SS 304	SS 316	20Cb3	825
0.05	60	10	12	—	—	—
0.1			25	—	—	—
0.2			62	—	—	—
20	102	3	—	—	—	41
38	110	2	—	2000	—	—
38	B.P.	4	—	—	—	10
48	B.P.	4	—	—	—	9
50	60	35	—	—	—	2
70	21	42	—	49	15	14
70	60	35	—	—	—	5
90	21	1	30	—	—	—
98	24–44	4.5	—	2	—	—

Table 4.4 Corrosion Rate of Various Austenitic Stainless Steels in Hydrofluoric Acid (mpy) Duplicate Samples Exposed for Five 48-Hour Test Periods, No Aeration

HF Conc. (Molar)	Temp (°C)	Type 201 (UNS S20100)	Type 304 (UNS S30400)	Type 316 (UNS S31600)	AL 29-4C (UNS S44735)	Al-6XN® (UNS N08367)
0.01	20	0.01	0.02	0.03	0.01	0.01
0.01	60	0.1	0.11	0.30	0.02	0.03
1.0	20	69.18	23.25	20.85	0.05	0.85
1.0	60	—	—	—	178.52	21.23
5.0	20	—	—	—	403.42	7.03

Also, cold-worked stainless steels (particularly Types 301 through 304) can fail by hydrogen embrittlement of the α-martensite phase that is formed. Type 316 and higher nickel alloys are resistant because of greater austenite stability.

Welds, including autogeneous welds, in Alloy 20 and in Alloy 825, can be attacked as a result of the columbium (niobium) in Alloy 20 and, to a lesser extent, of the titanium in Alloy 825. Only seamless tubes should be used in severe services.

Nickel and Nickel-Based Alloys

Nickel and nickel-based alloys have a range of corrosion resistance and usefulness in aqueous HF. Figure 4.2 shows that nickel alloys are useful across the range of HF

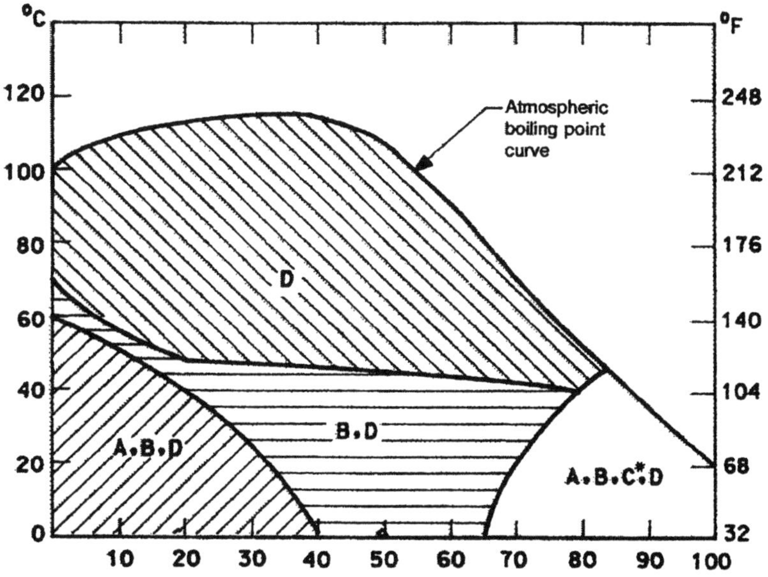

Figure 4.2 Metals and Alloys for Hydrofluoric Acid

Iso-corrosion Regions Where Observed Corrosion Rates Are 0.5 mm/y (20 mpy) or Less

A N02200, N06030, N06600, N06985, N08007, N08020, N08825
B N06022, N10276, N10665
C Carbon steel*
D C70600, C71500, N04400, N24135, P00020, P04995, P07015, R03600

Refer to text on nickel-rich and nickel-based alloys for information on stress corrosion cracking (SCC). This information is for guidance only. It represents low-flow, oxygen-free, uncontaminated conditions. Velocity and/or impurities may make these selections unsuitable.
*May suffer HSCC, hydrogen embrittlement (HE), or hydrogen-induced cracking (HIC).

concentrations up to the indicated temperatures. Oxygen in HF increases corrosion of these alloys. The presence of oxidizing salts in HF may be beneficial to nickel-chromium alloys and detrimental to nickel-copper and nickel-molybdenum alloys.

Morphology of Corrosion Attack in Aqueous HF Solutions. Rebak[6] studied the corrosion behavior of nickel alloys in HF and made some interesting observations on the nature of the corrosive attack. Unlike other acids such as sulfuric and hydrochloric, the general corrosion of nickel alloys in hydrofluoric acid is not uniform. Rebak[6,7] observed that in HF, the tested coupons seemed more attacked in certain areas than in others and this uneven mode of corrosion was highly dependent on the composition of the alloy, temperature, acid concentration, if the wet acid is in the liquid or vapor state, and on the level of oxygen that was present. It was commonly observed that the corrosion attack was cavernous in nature. Hydrofluoric acid also tended to dissolve and re-precipitate some elements of the alloy preferentially. Sometimes, corroded coupons appeared to have been plated with either nickel or copper or both. The color appearance of the coupons was different for each alloy under each set of test conditions and, even for the same alloy, under different test conditions. The mode of internal attack was also different for each alloy. The modes described in the study included uneven corrosion or serration of the surface, dealloying, intergranular attack, lace-like or cobweb penetration, internal voids, pits or cavities, wasp nest appearance and long and thin transgranular cracks.

Chromium-Free Alloys The nickel-copper alloys, typified by Alloy 400 and modifications of this composition K-500, exhibit excellent corrosion resistance to hydrofluoric acid over a wide range of acid concentrations and temperature. The isocorrosion diagram for Alloy 400 is shown in Figure 4.3.[4] The corrosion resistance of Alloy 400 is less affected by acid concentration, temperature, or velocity than it is by the amount of oxygen in contact with the solution. Data generated by Pawel[8] on a number of nickel alloys is presented in Tables 4.5. The data at various temperatures shows that the presence of air in HF vapors dramatically increass the corrosion rates. For example, in a 10% HF solution at 50°C the corrosion rate of Alloy 400 is an acceptable 0.15 mm/y in the immersed phase, whereas in the air laden vapor phase the corrosion rate was determined to be 5.87 mm/y. Further discussion and data for Alloy 400 in comparison with other Ni-Cr-Mo alloys are presented in the section below.

Welds and Castings. Weld deposits that contain niobium suffer accelerated corrosion in aqueous HF.[4] For welding Alloy 400, titanium, which reduces the loss of corrosion resistance considerably less than does niobium, has been substituted in welding filler metals. Corrosion data for cast alloys M-35-1 (N024135) and M-30C (N024130) and weld materials with various amounts of niobium are presented in Table 4.6.[9] The cast alloy N 024130 with less than 0.5%Nb had lower corrosion than N024130 (1.3% Nb) in 49% HF. Weld deposits with 2.08% Nb had higher corrosion in aerated and nitrogen purged HF than those without Nb.[4] Welds in alloy 400 made with gas tungsten arc (GTAW) with ERNiCu (N04060) and shielded metal arc weld (SMAW) ENiCu (W84190) processes were selectively corroded in 5-15% HF with

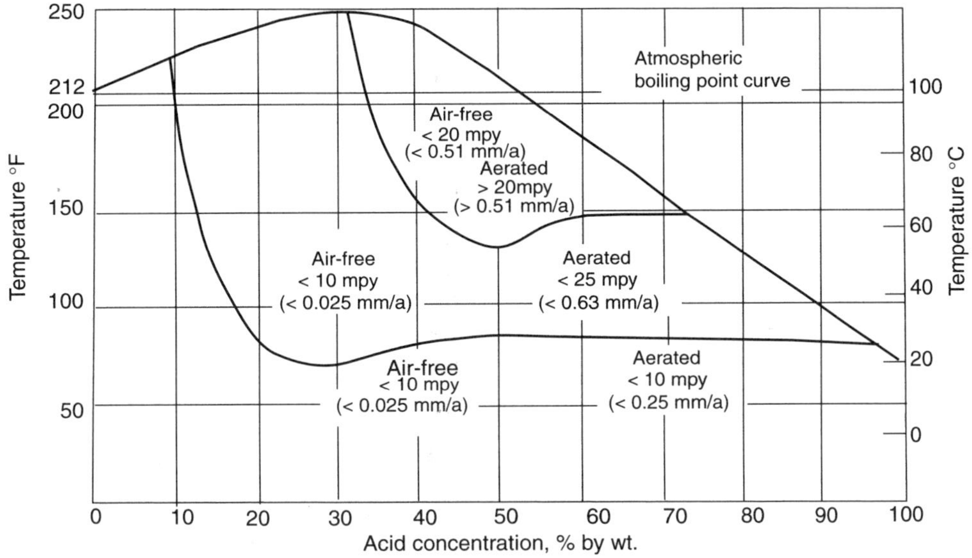

Figure 4.3 Isocorrosion Chart for UNS N04400 in Hydrofluoric Acid Air Free and Aeration

Table 4.5 Corrosion Rates (mm/y) of Nickel Alloys in 1% and 10% HF Immersion Tests[A]

Alloy	HF Concentration	24°C		50°C		76°C	
		Immersed	Vapor	Immersed	Vapor	Immersed	Vapor
Alloy 200	1.0%	0.06	1.11#	0.02	1.85#*	0.18	3.53#*
	10.0%	0.07	1.87#	—	—	0.27	4.47#*
Alloy 400	1.0%	0.04	0.91	0.18	3.38	0.21	2.27
	10.0%	0.05	2.24#	0.15	5.87#	0.30	1.39
Alloy 600	1.0%	0.05	1.06	0.33	3.36*	0.22	2.59*
	10.0%	0.07	8.36#	0.36	10.7*	0.33	5.00*
Alloy 22	1.0%	0.01	0.005	0.05	0.04	0.19	0.59
	10.0%	0.03	0.09	0.19	1.25	0.39	2.33
Alloy C-276	1.0%	0.02	0.04	0.12	0.32	0.21	1.25
	10.0%	0.06	0.35	0.21	2.35	0.34	2.57*

[A]Rates are calculated assuming uniform weight loss.
#Indicates extreme attack at vapor-solution interface.
*Adherent corrosion products on sample leading to underestimation of corrosion rate.

Table 4.6 Corrosion Rates of Ni-Cu Alloys Showing the Effects of Niobium Corrosion Rates mm/y (mpy)

Alloy (Nb)	49% HF 49°C (120°F) 120 hrs	40% HF, N_2 Purge, 70°C (158°F)16.5 d	40% HF, Aerated, 70°C (158°F)16.5 d	40% HF, N_2 Purge, 70°C (158°F) 12 d	40% HF, Aerated, 70°C (158°F) 12 d
M35-1 (0.5%)	0.35 (14)				
M30C (1.3%)	0.62 (24.5)*				
Weld Filler (2.08%)		0.09 (3.6)	1.7 (67)	0.05 (2)	1.07 (42)
Weld Filler (2.08%)		0.71 (28)	2.64 (104)	0.05 (2)	1.32 (52)*

*Localized attack.

sulfuric acid. Weld repairs and restoration were made with C71500 welding consumables ECuNi (W60715) and ERCuNi (C71580) which showed no attack.

Alloy 200 is less resistant to corrosion in aqueous HF than Alloy 400 or 600. Its use is limited to completely air-free systems below 80°C (175°F). It is, however, one of the more resistant metals to hot gaseous hydrogen fluoride, as discussed in Chapters 3 and 5.

Alloy B-2 has useful resistance to aqueous HF but is not used because of cost and the fact that it offers no advantage over alloys of nickel and copper. Actually, B-2 is not as resistant as Alloy 400 and is likewise subject to accelerated corrosion in the presence of oxidizing contaminants. The iso-corrosion chart for alloy B-2 is shown in Figure 4.4.

Comparative corrosion data from various sources for Alloy 400 and B-2 is summarized in Table 4.7. The data shows that in 70% HF welded Alloy 400 and Alloy B-2 are equally resistant, with a corrosion rate of 2 mpy or less, whether or not the solutions are aerated.[11]

It has been noted by Sridhar that the corrosivity of HF is quite different from HCl and HBr. "For example, in HCl environments the corrosion resistance of austenitic stainless steels and nickel-based alloys is *enhanced* by the presence of Mo and W and *decreased* by chromium. In HF, similar behavior is not observed. Instead, the corrosion resistance is strongly dependent on the nickel content. . . ."[13]

Based on such corrosion data and experience, the "high-performance" nickel alloys, B-2, although having excellent resistance to AHF and high-temperature HF gas, are seldom used in aqueous HF service because lower alloys perform as well and at lower cost.

Chromium-Bearing Alloys The Ni-Cr-Fe and Ni-Cr-Mo grades behave quite differently in many instances. Crum[14] using data from the literature and unpublished data presented the results in an iso-corrosion, temperature-concentration format. The compilation of data is presented in Figures 4.5 to 4.16. Where possible an isocorrosion

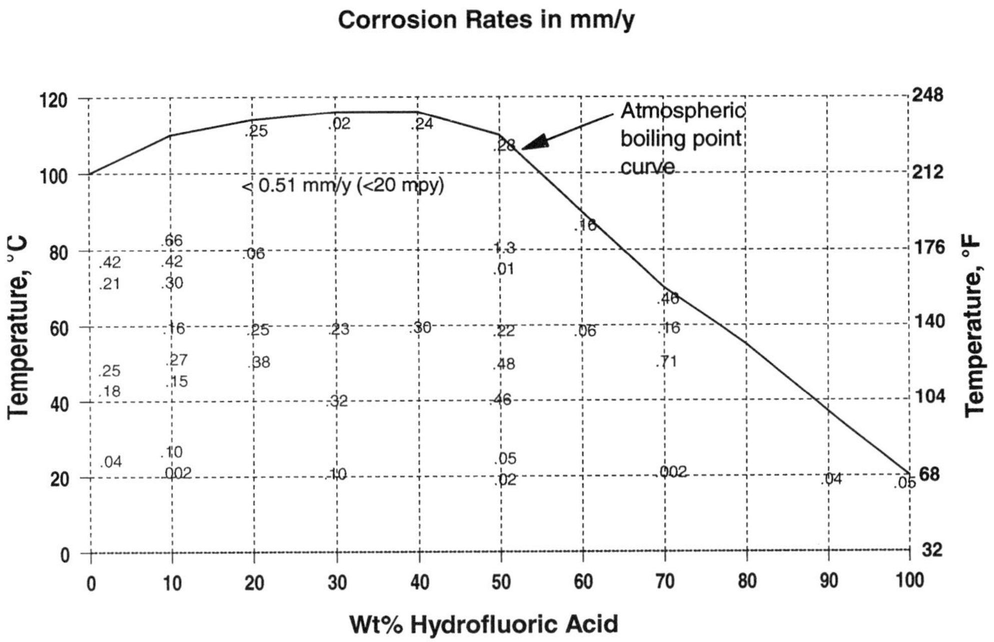

Figure 4.4 Hydrofluoric Acid Corrosion Data for Alloy B-2 (UNS N10665)

Table 4.7 Corrosion Rate of Alloy 400 and B-2 in Various HF Solutions

Environment	Corrosion Rate (mpy) Alloy 400 (N04400)	Alloy B-2 (N10665)
70% HF at 52°C, *aerated* cast Alloy 400 (M-35-1) coupled to INCO 60, Butt weld, 336 hours[11]	2*	
70% HF at 52°C, *aerated* cast Alloy B-2 (N-4M) coupled to Alloy B-2, Butt weld1[11]		< 1*
70% HF at 52°C, *deaerated* cast Alloy 400 (M-35-1) coupled to INCO 60, Butt weld[11]	2	
70% HF at 52°C, *deaerated* cast Alloy B-2 (N-4M) coupled to Alloy B-2, Butt weld		2
48% HF, 52°C, aerated[11]	2/2	190/160
5% HF, 70°C [12]	13	15
48% HF, 79°C	<1	25

*Lower corrosion rate than expected, due to ferrous corrosion products from other test samples.

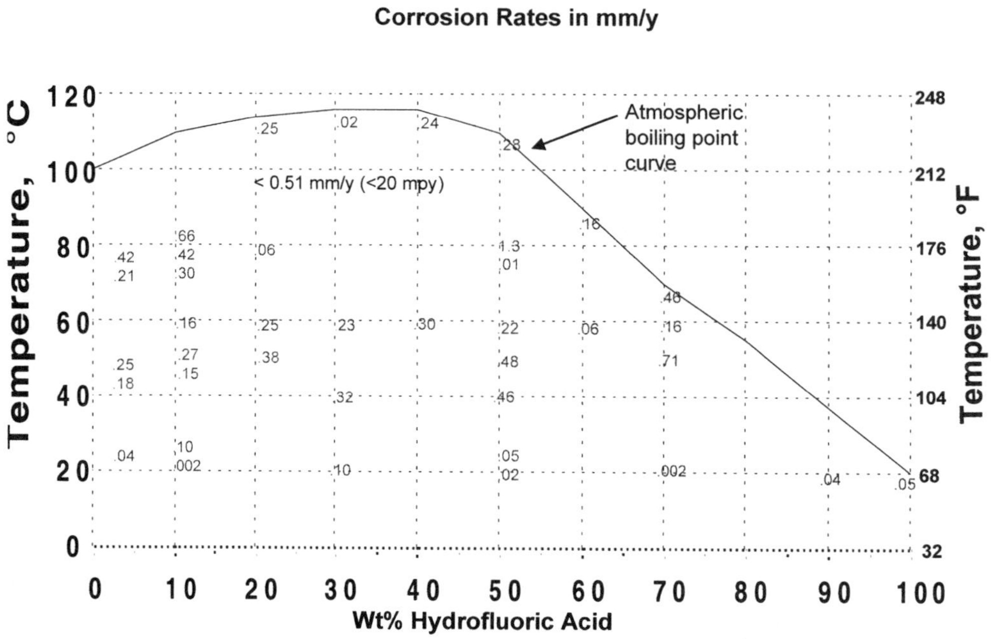

Figure 4.5 Hydrofluoric Acid Corrosion Data for Alloy 400 (UNS N04400)

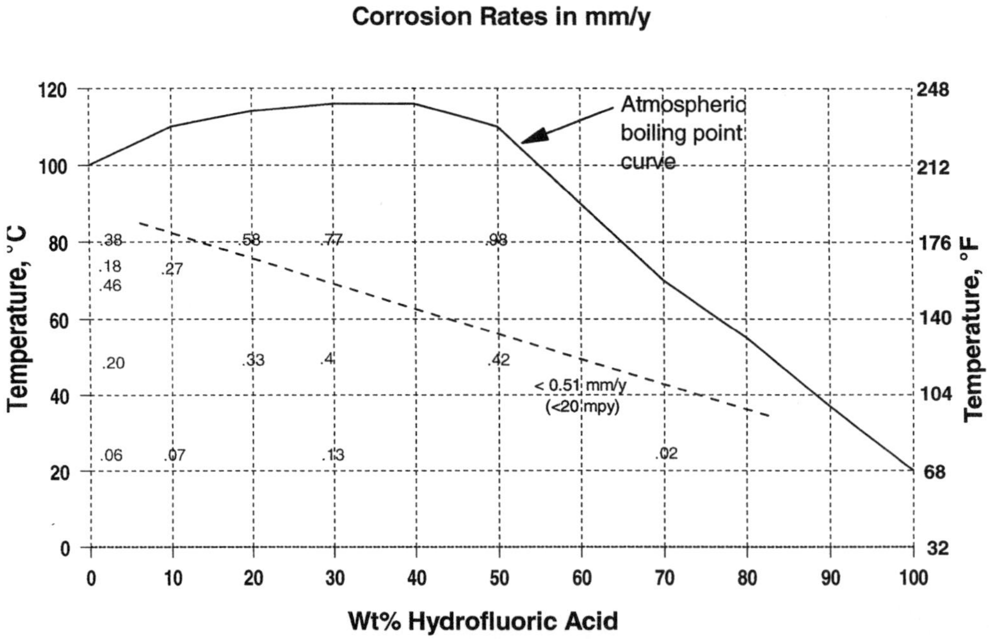

Figure 4.6 Hydrofluoric Acid Corrosion Data for Alloy 200 (UNS N02200)

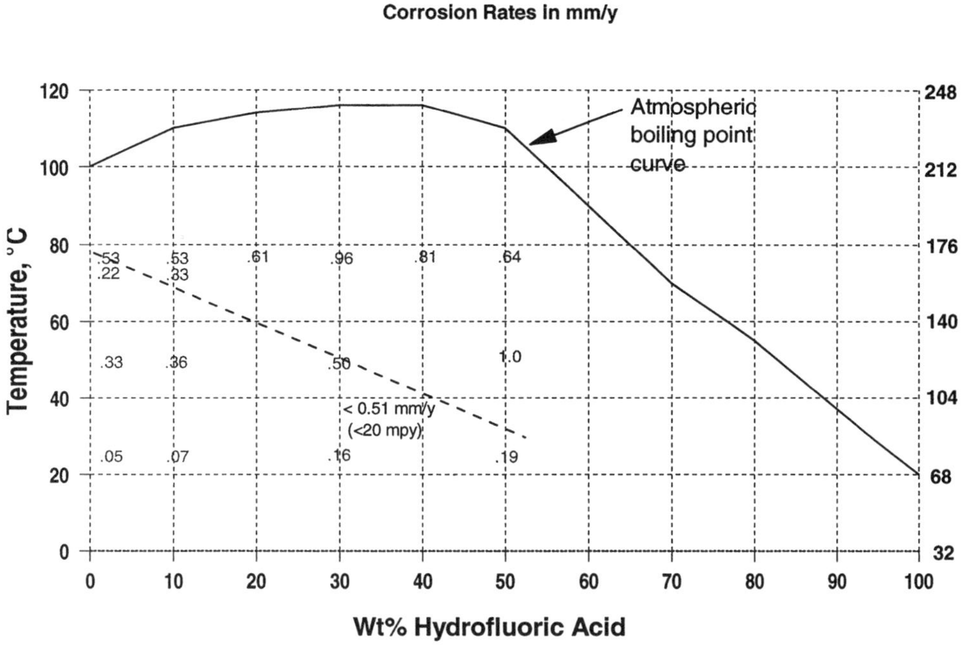

Figure 4.7 Hydrofluoric Acid Corrosion Data for Alloy 600 (UNS N06600)

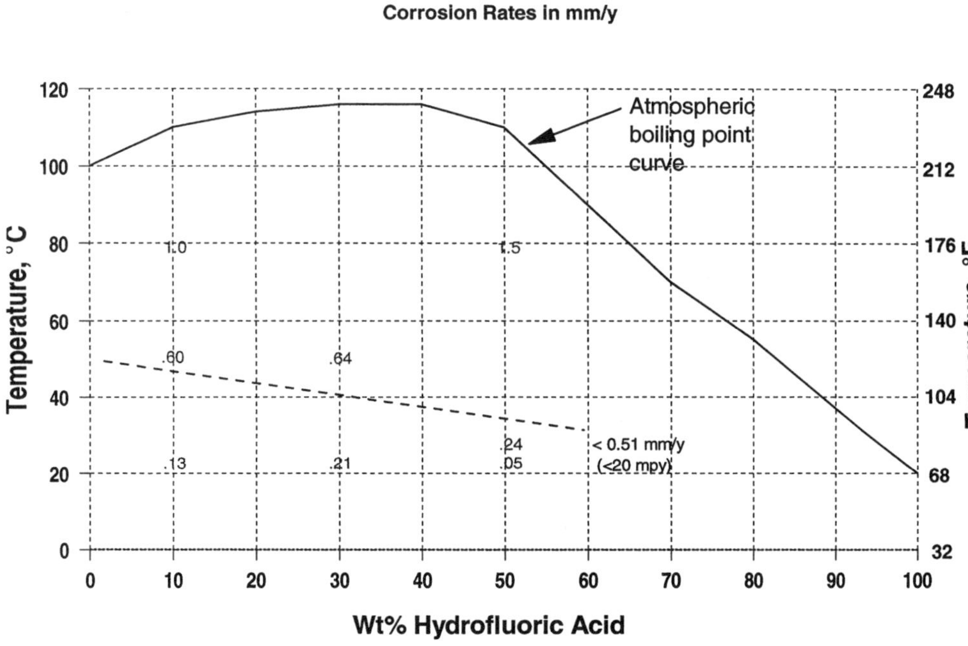

Figure 4.8 Hydrofluoric Acid Corrosion Data for Alloy 690 (UNS N06690)

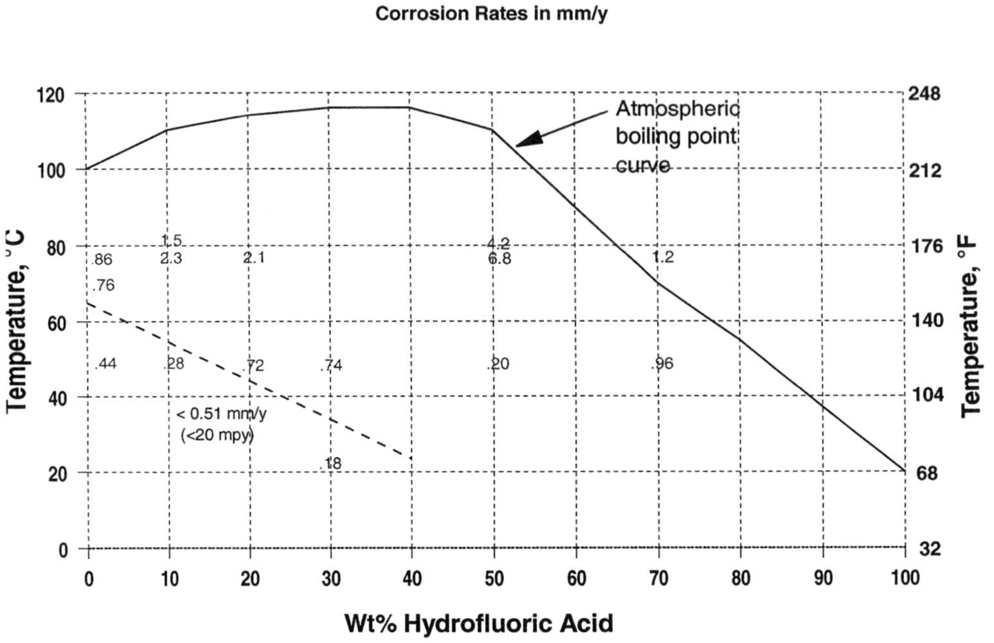

Figure 4.9 Hydrofluoric Acid Corrosion Data for Alloy G-30 (UNS N06030)

line representing 0.5 mm/y (20 mpy) is also drawn on the charts. The data presented in these charts is for HF under air free conditions.

Alloy 600 Alloy 600 is resistant to dilute aqueous HF solutions at ambient temperatures but may be corroded at appreciable rates in hot aqueous acid. The presence of iron salts, presumably ferric ions, improves the resistance of Alloy 600.[4] A chart plotting the corrosion rate of Alloy 600 is shown in Figure 4.6.[14]

It should be noted that the commonly used filler metals for welding Alloy 600 (i.e., 62, 82, and 182) contain appreciable amounts of columbium, which can be a problem at high temperature.

Schmitt and Losacker[15] evaluated a number of chromium-bearing Fe-Ni-Cr-Mo alloys in HF at low temperatures. Figures 4.17 and 4.18 show the data obtained for a number of alloys in aerated HF and nitrogen deaerated HF. There is a significant reduction in corrosion in nitrogen purged 20% and 40% HF for Alloy G-30 and a 6% molybdenum alloy, 1925hMo. The data also showed that in aerated HF solutions, the corrosion resistance of these alloys benefits primarily from chromium and molybdenum content with nickel adding some limited positive effects. In deaerated HF solutions, positive benefits in corrosion resistance result for alloys with nickel, molybdenum, and copper (beneficial at 100mV$_H$). The combination of nickel and copper are attributed to low corrosion rates of 1925hMo and G-30.

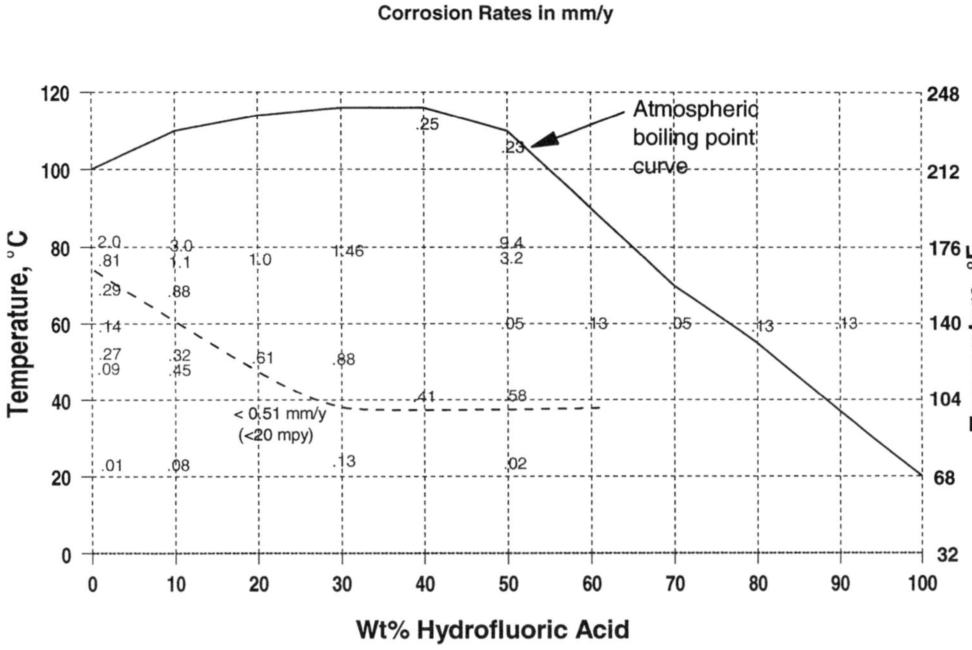

Figure 4.10 Hydrofluoric Acid Corrosion Data for Alloy 825 (UNS N08825)

Ni-Cr-Mo High-Performance Alloys The Ni-Cr-Mo alloys do well in certain temperature and concentration ranges because of their high nickel content. Their high molybdenum (and sometimes tungsten) contents, although mildly beneficial, do not provide the dramatic improvement in corrosion resistance that they do in some other acid environments.

As noted previously above, Alloy C-276 and its variants or derivatives are not as resistant as the higher nickel Alloys 400 or 600. There are exceptions, of course, as in solutions formed in the reprocessing of nuclear fuel elements in hot 8M (16%) HF, where Alloy C-276 was the best material tested, despite its high corrosion rate.[16]

Pawel[8] compared the corrosion rates of wrought and welded samples of Alloy 22 and Alloy 400 in 24% and 48% HF concentrations. The results of the study are summarized in Table 4.8. As expected, the data shows an increase in corrosion rate with temperature and concentration. The wrought and welded Alloy C-22 samples when exposed to HF vapors in air had lower rates than Alloy 400 for similar consitions. However, Alloy 400 in immersion conditions had lower rates than Alloy C-22. Even though welded Alloy 400 had generally higher rates than Alloy 400 samples, these rates were generally lower than Alloy C-22 welds under similar conditions.

The various iso-corrosion diagrams presented in Figures 4.5–4.16 are summarized in Figures 4.19 and 4.20. From this figure it can be concluded that Alloy 400 has acceptable rates up to the boiling point at all concentrations. Notable among the Ni-

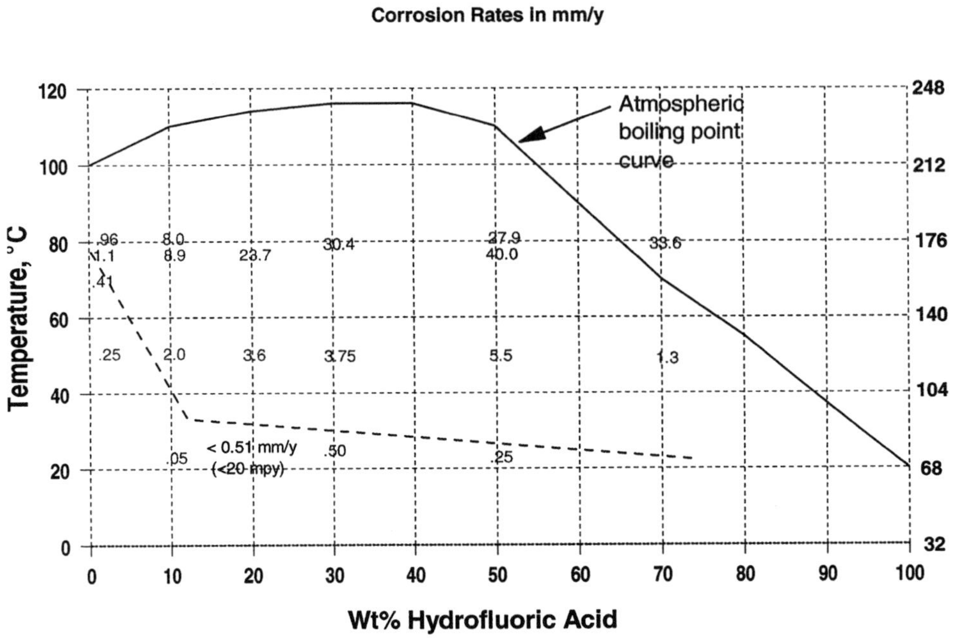

Figure 4.11 Hydrofluoric Acid Corrosion Data for Alloy 625 (UNS N06625)

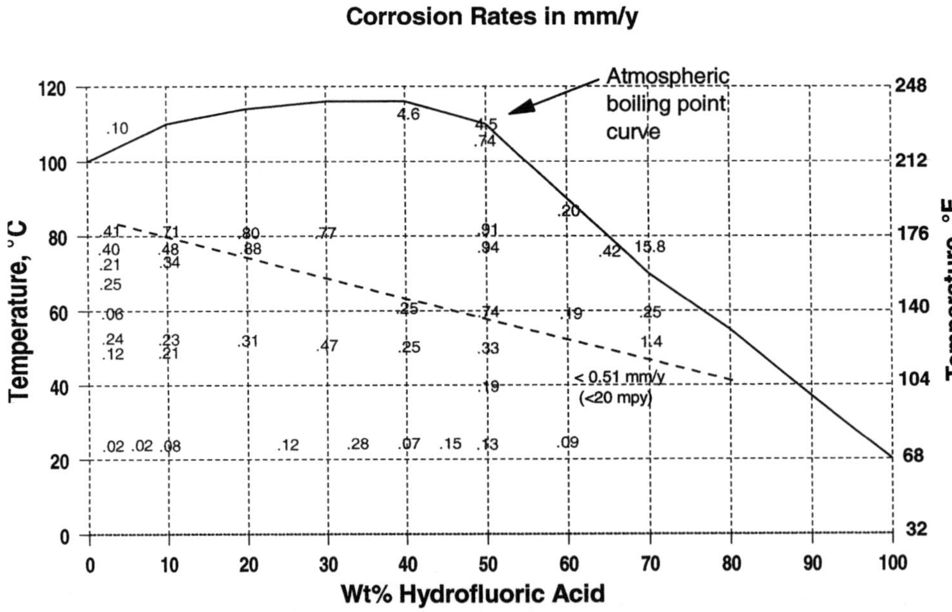

Figure 4.12 Hydrofluoric Acid Corrosion Data for Alloy C-276 (UNS N10276)

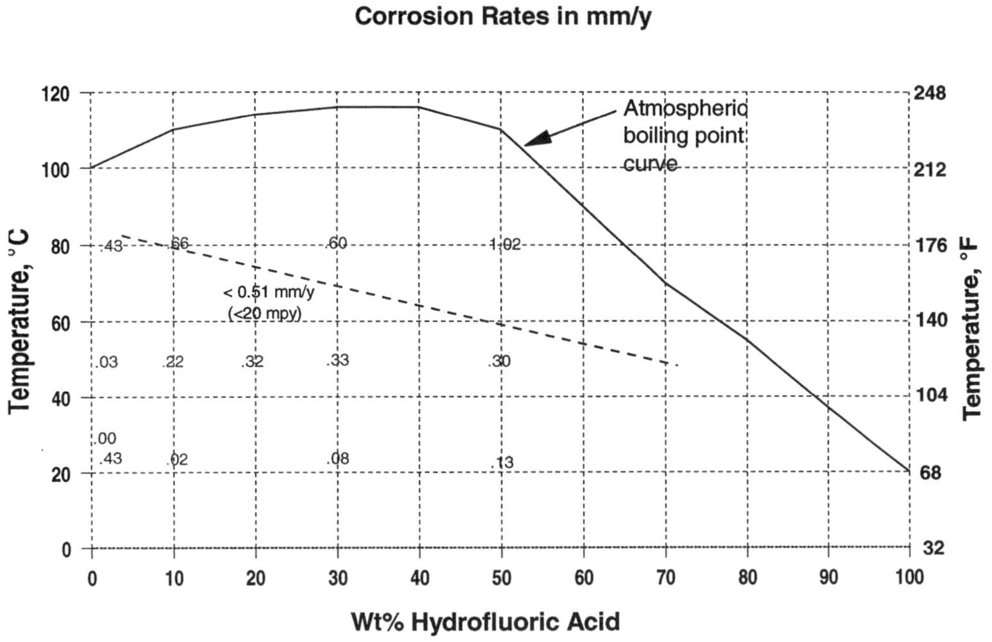

Figure 4.13 Hydrofluoric Acid Corrosion Data for Alloy 686 (UNS N6686)

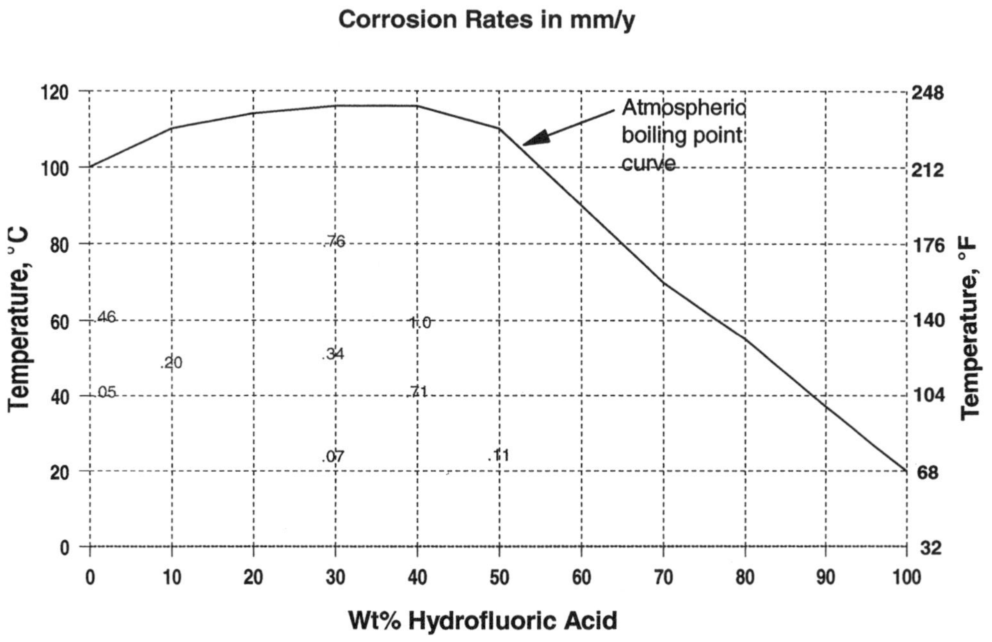

Figure 4.14 Hydrofluoric Acid Corrosion Data for Alloy 59 (UNS N06059)

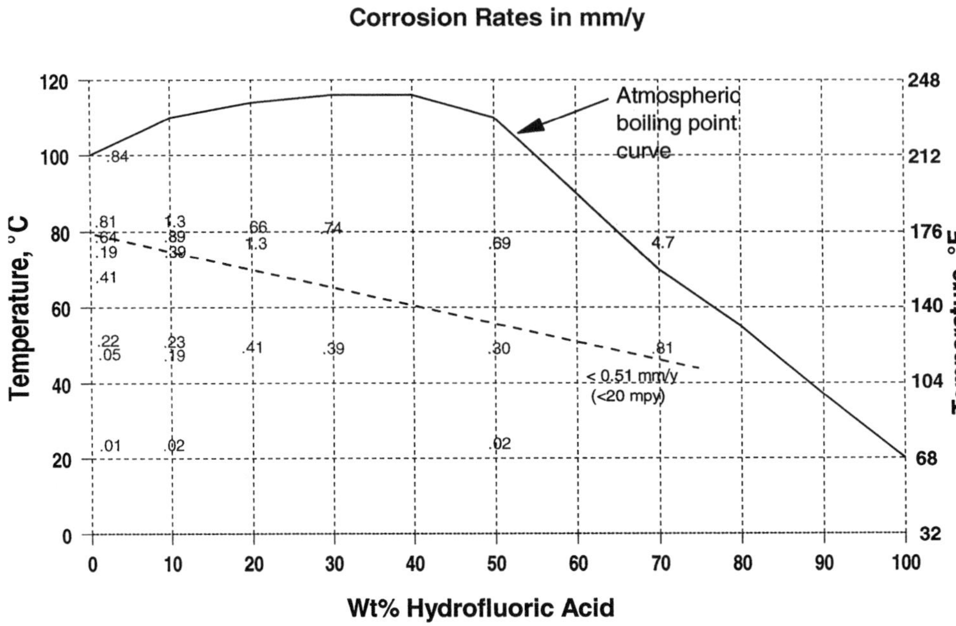

Figure 4.15 Hydrofluoric Acid Corrosion Data for Alloy C-22 (UNS N06022)

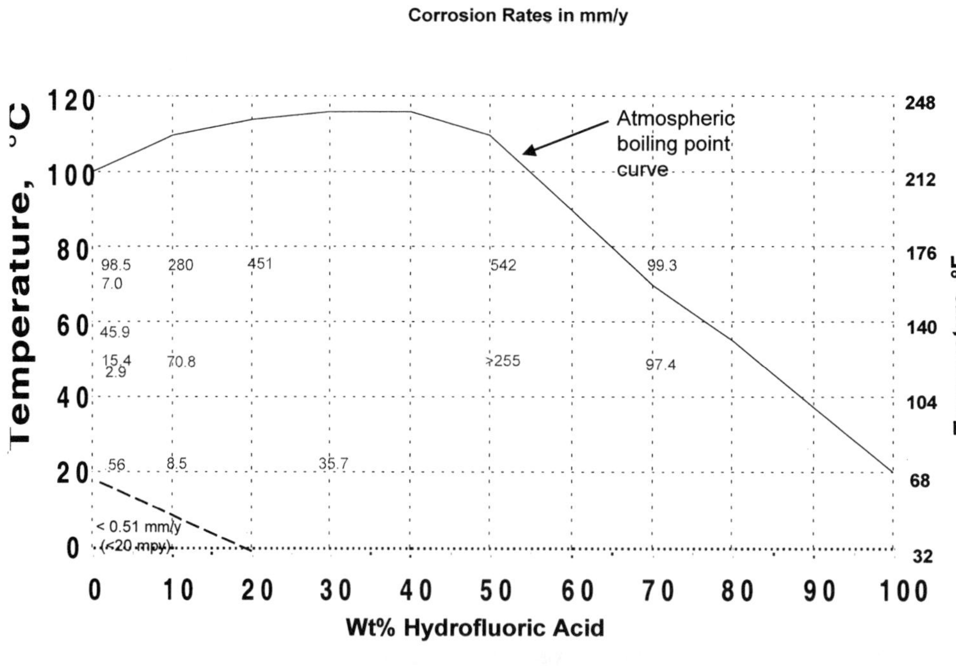

Figure 4.16 Hydrofluoric Acid Corrosion Data for Alloy 316L (UNS S31603)

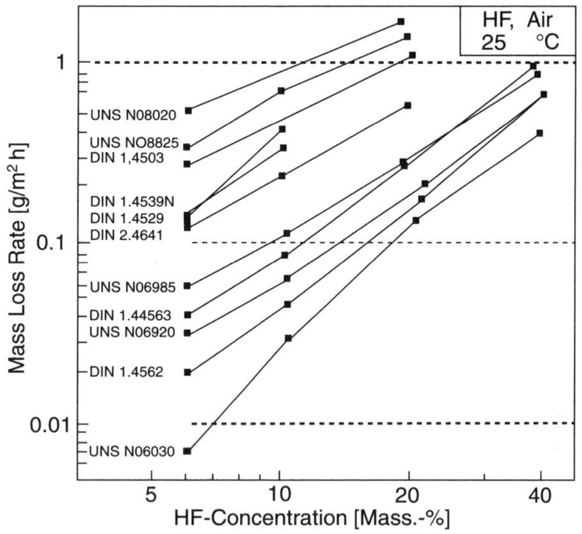

Figure 4.17 Corrosion of Various Alloys in HF with Air at 25° C

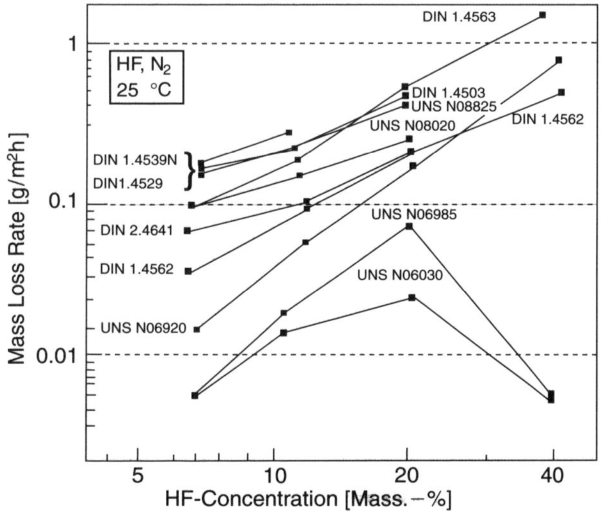

Figure 4.18 Corrosion of Various Alloys in HF in Nitrogen at 25° C

Cr-Mo alloys is the lower useful conditions for Alloy 625. The lower corrosion resistance of Alloy 625 is thought to be due to the 3.5% Nb content.

Rebak[7] studied the behavior of a copper containing Ni-Cr-Mo Alloy C-2000 and Alloy 400 in HF concentrations up to 20%. The study showed that the corrosion rate of both alloys increased with the acid concentration and temperature, The corrosion

Table 4.8 Corrosion Rates (mm/y) of Alloy C-22 and Alloy 400 in 24% and 48% HF Immersion Tests

Alloy	HF Concentration	24°C Immersed	24°C Vapor	50°C Immersed	50°C Vapor	76°C Immersed	76°C Vapor
Alloy 22	24%	0.06	0.48	0.29	1.51	1.42	3.81
	48%	0.06	0.50	0.19	0.95	0.43	3.81
Welded Alloy 22	24%	0.38	0.43	0.68	1.90	1.63	3.30
	48%	0.34	0.60	1.65	1.14	3.00	2.34
Alloy 400	24%	0.08	1.67	0.28	5.26	0.36	5.21
	48%	0.06	1.94	0.34	6.17	0.44	5.61
Welded Alloy 400	24%	0.08	1.55	0.34	5.72	0.36	5.21
	48%	0.44	2.02	0.33	5.41	0.69	5.28

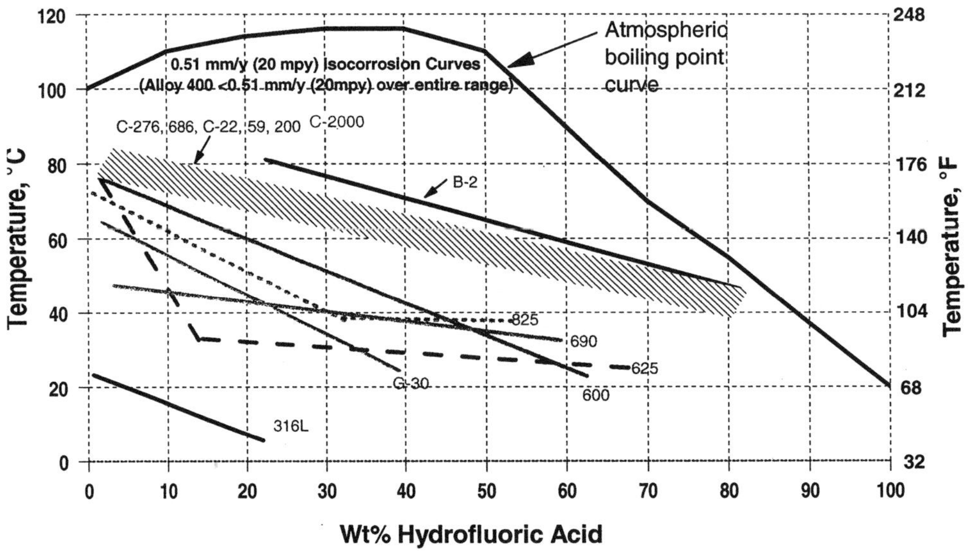

Figure 4.19 Summary Hydrofluoric Acid Isocorrosion Chart Showing 0.51 mm/y (20 mpy) Lines for the Various Alloys Examined

rate of Alloy 400 increased faster than the corrosion rate of C-2000 alloy, especially in the vapor phase. Figures 4.21 to 4.24 summarize the data obtained. Alloy 59 is a similar alloy to C-2000 in that it shows good corrosion resistance over a wide range of HF concentrations. The data is summarized in the iso-corrosion curve in Figure 4.14.

Crum[14] also studied the behavior of several alloys under condensing conditions at 50°C with 5% HF, 8% O_2, 15% H_2O, and balance nitrogen and carbon dioxide. The data showed that there was no clear effect of alloying elements but the high Cu and

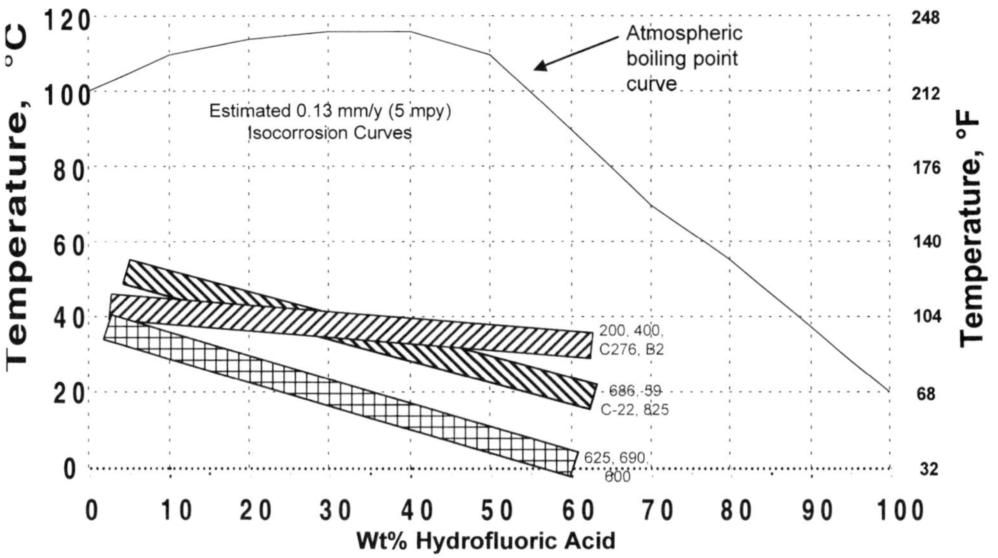

Figure 4.20 Summary Hydrofluoric Acid Isocorrsoion Chart Showing 0.13 mm/y (5 mpy) Lines for the Various Alloys Examined

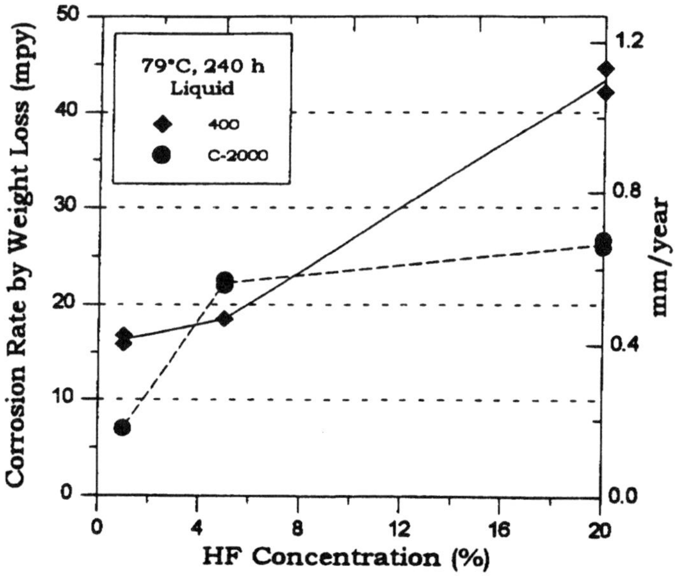

Figure 4.21 Effect of Acid Concentration on the Corrosion Rate in the Liquid Phase

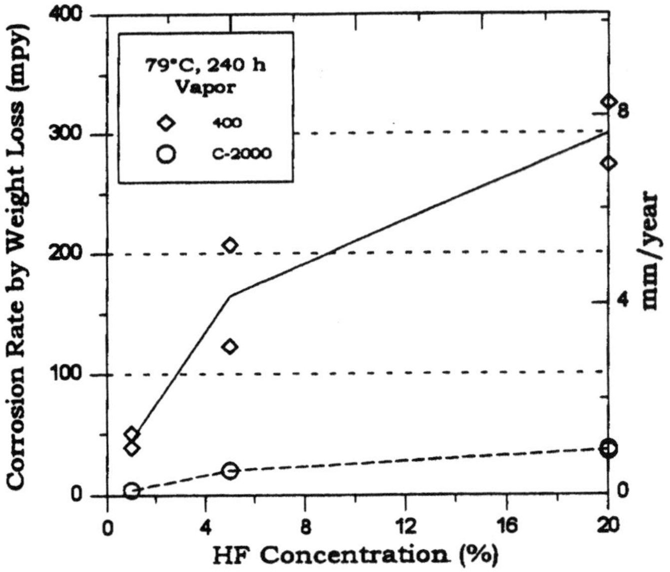

Figure 4.22 Effect of Acid Concentration on the Corrosion Rate in the Vapor Phase

Fe alloys—Alloy 400 and 25-6-Mo—fared the worst. The Ni-Cr-Fe and Ni-Cr-Mo alloys had good, very good to moderate corrosion rates. The data is summarized in Table 4.9.

Summary The relative corrosion resistance of several alloy groups in the liquid phase with no aeration from best to worst may be summarized as follows:[14]

Ni-Cu > Ni-Mo > Ni-Cr-Mo > Ni-Cr-Fe Ni-Cr-Fe-Mo >> Fe-Cr-Ni-Mo

For liquid phase exposure, alloying nickel with copper is most beneficial while alloying with molybdenum is the second most effective. Combined additions of Cr, Mo, and W are beneficial and better than additions of Cr and Fe or Cr, Fe and Mo. Addition of niobium is detrimental. Although Alloy 400 is susceptible to vapor phase stress corrosion cracking, it is more resistant to aqueous HF than most other materials at all concentrations up to the boiling point and is more economical than many other alloys. Of the Ni-Cr-Mo alloys, Alloy 2000 has the best corrosion resistance and needs to be given consideration in both liquid and vapor phase HF solutions.

Nickel and Nickel Alloy Stress Corrosion Cracking

Alloy 400 and K-500 Alloy 400 and Alloy K-500 are susceptible to intergranular stress corrosion cracking (SCC) when exposed to moist vapors of HF in the presence

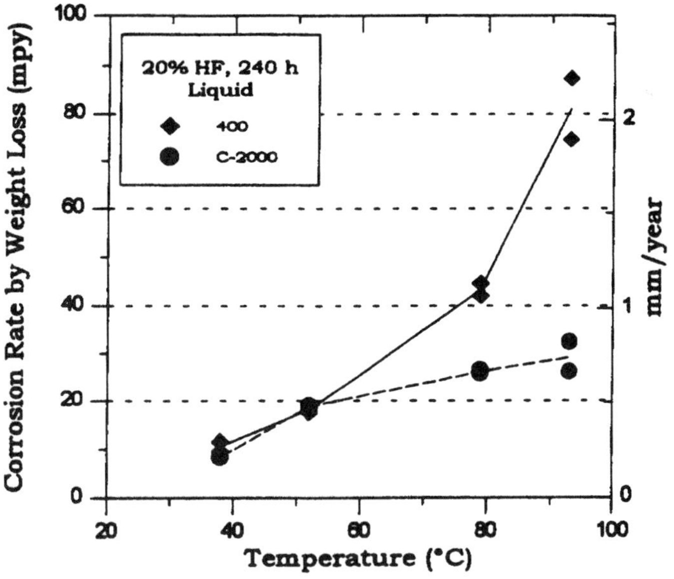

Figure 4.23 Effect of the Temperature on the Corrosion Rate in the Liquid Phase

of air. The intergranular stress corrosion cracking of a cold bent instrument tube from an HF refining column has been described.[17] The tube was exposed to moist vapors during a shutdown when the nitrogen purge was inadvertently turned off. The column, constructed of Alloy 400 and post-weld heat treated (PWHT), failed from SCC after several years of service and was replaced with a more resistant alloy.

Schussler[18] studied welded Alloy 400 after exposure to 5% and 48% HF. SCC was observed in vapor exposed samples. It was noted that oxygen saturation in the vapor space and the level of stress played a role in the severity of SCC. No SCC was found on submerged samples. Cracking was found in the base metal and weld metal of as-welded samples. No cracking was observed on wrought solution annealed samples, and in addition no cracking was observed on welded and stress relieved samples that were welded with gas tungsten arc (GTAW) process. Full anneal, but at a minimum, stress relief of Alloy 400 after welding was recommended to avoid SCC.

Copson and Cheng[19] investigated SCC of a number of nickel alloys in HF. The data is shown in Table 4.10. Intergranular and transgranular cracking of Alloy 400 and Alloy K-500 in 48% HF vapor at 60°C was observed.

Although it has been long known that Alloy 400 can fail by IGSCC in HF, the mechanism was not understood until it was demonstrated that aqueous HF containing cupric fluoride (CuF_2) could cause this type of failure.[20] As can be seen from Figure 4.25, the least resistant nickel-copper composition is that of Alloy 400. It should be noted, from examination of the graph, that cupronickel alloys with up to 30% nickel are more resistant to this phenomenon. The data also showed that increasing the concentration of CuF_2 in HF reduces time for SCC, and increasing tensile

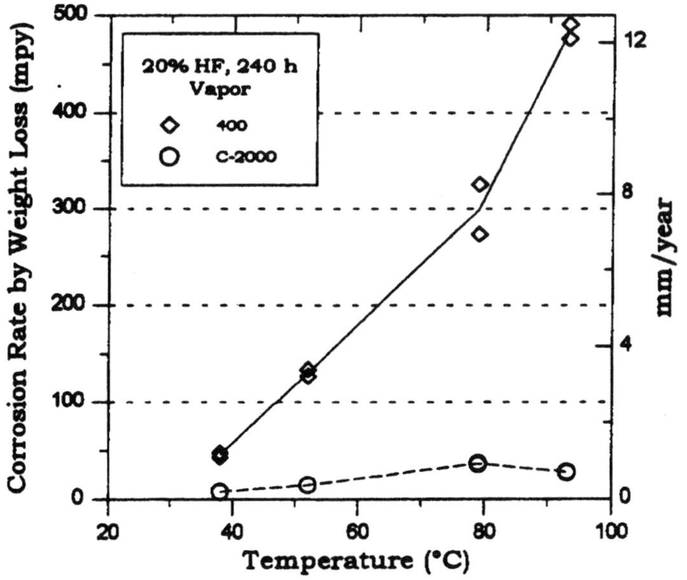

Figure 4.24 Effect of the Temperature on the Corrosion Rate in the Vapor Phase

Table 4.9 Corrosion Rate of Various Alloys Tested in HF Gas Stream Under Condensing Conditions

	Corrosion Rate, mm/y (mpy)	
Material	70°C, 100 hours	50°C, 55 hours
Alloy 600	0.02 (1.0)	0.23 (9.0)
Alloy 400	0.15 (6.1)	1.90 (74.8)
Alloy 825	0.04 (1.6)	0.62 (24.5)
Alloy C-276	0.03 (1.1)	0.39 (15.3)
Alloy 686	0	0.15 (6.0)
Alloy 622	0.02 (0.9)	0.66 (26.2)
Alloy 25-6Mo	0.09 (3.5)	0.94 (37.1)

stresses reduced time for SCC. SCC has not been a problem in liquid streams because a sufficiently high concentration of CuF_2 probably cannot be attained. It occurs in the vapor state, due to exposure to moist vapors and an accumulation of CuF_2 on the metal surface. Pipelines and equipment should be blanketed with *nitrogen* rather than air when drained or when otherwise empty or only partially full.

Thermal stress relief has not proven entirely successful in preventing SCC. As can be seen from Figure 4.26, SCC can occur at relatively low stress levels. However, stress relief will prolong life expectancy. It is recommended that stress relief be performed in the temperature range 675–730°C (1250–1350°F). The vessel should then preferably be allowed to cool in the furnace or, alternatively, in still air.

Table 4.10 Results from Nickel Alloy U-bend in 48% HF Vapors at 60°C (140°F)

Alloy	Condition	Hardness VHN 30pd	Hardness HRc	First Observed Cracking, Days
Alloy K-500	Hot Rolled	148–210		2–6
Alloy K-500	Aged		32	5
Alloy 400	Annealed	109–136		4–15
Alloy 400	Cold Rolled	250		6–15
Alloy 400	Welded			4–14
Alloy 400 + 1% Si	Cold Rolled	293		15
85Ni-15Cu	Annealed	147		5
Alloy 200	Cold Rolled	232		14
Alloy 600	Cold Rolled	295		14–15

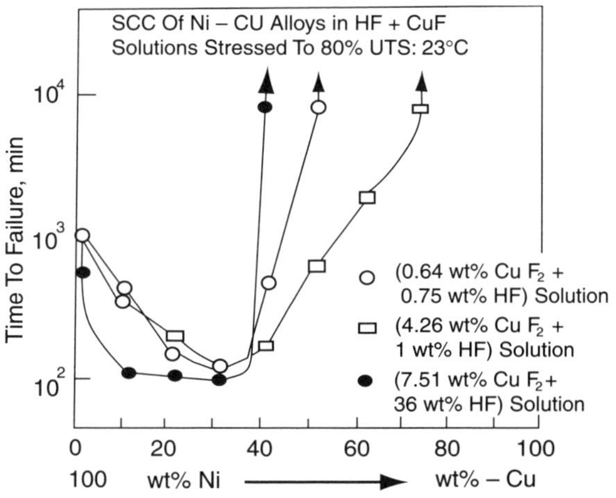

Figure 4.25 Plot of Time for Stress Corrosion Cracking of Various Nickel-Copper Alloys, Exposed to HF, with CuF_2 at 23°C while Stressed at 80% of Ultimate Tensile Strength

Alloy 200 From Table 4.10 it is notable that cold rolled Alloy 200 suffered from SCC in vapors of 48% HF at 60°C (140°F). Deep intergranular attack on the tension side of the u-bends was reported.[19] These cracks opened up with additional bending. No mechanistic explanation has been provided for the cracking of Alloy 200. It has been reported that welded Nickel 200 in AHF service suffered SCC due to exposure to CuF_2 that may have been transferred from copper-bearing alloys up stream. [21]

Alloy 600 Transgranular cracking of Alloy 600 in vapors of 48% HF at 60°C (140°F) is also reported in Table 4.10. Considerable general corrosion and pitting was also

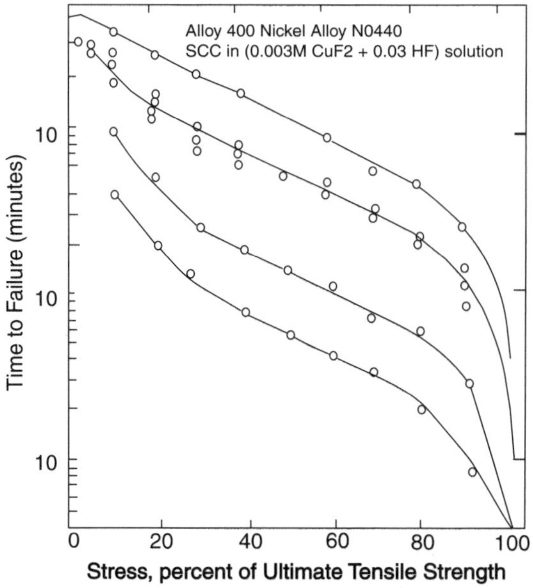

Figure 4.26 Effects of Stress and Temperature on SCC of Alloy 400 in $HF + CuF_2$ Solutions

observed on the samples. Users of Alloy 600 have reported SCC in HF or AHF when CuF_2 is present.[21] In-service cracking of Alloy 400 has occurred in AHF at 135°C (275°F), and with 5% Cl_2 at 100°C (212°F). In both cases, the environments caused only very low corrosion rates and transgranular cracking occurred within days of initial service. The cracks were found only in cold-formed dished heads.[10] No investigation was made into the cracking mechanism. Weld repairs were made without difficulty. In similar environments, annealed Alloy 600 had no cracking. It was concluded that to prevent cracking in AHF, Alloy 600 should be solution annealed after cold forming.

Ni-Cr-Mo Alloys The susceptibility of several nickel-based alloys to SCC or stress assisted internal penetration was investigated by using u-bend coupons.[6] Figures 4.27 and 4.28 show the average crack propagation rate for the various alloys exposed in 20% HF at 66°C and 93°C. At 66°C, the highest penetration (intergranular cracking) corresponded to Alloy 400 in the vapor phase, while the Ni-Cr-Mo Alloys C-2000 and C-276 were not susceptible to cracking. At 93°C the data showed Alloy 400 to be free from cracking in the liquid and vapor phase, while alloy C-276 was susceptible to cracking in both the liquid and vapor phase and Alloy C-2000 was found to be susceptible to the liquid phase. No explanation was provided for the difference in behavior of Alloy 400 at the two temperatures; however this may be related to the higher corrosion rate of the Alloy 400 coupons.

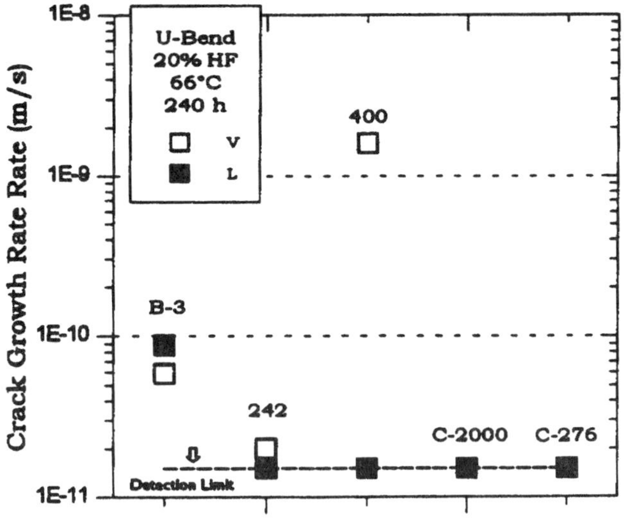

Figure 4.27 SCC Susceptibility of Nickel Alloys in 20% HF at 66°C. For C-2000 and C-276 Alloys the Symbols for the L and V Phases Overlap (no detected cracking).

Copper and Copper-Based Alloys

Any possible application of copper alloys in aqueous hydrofluoric acid is absolutely contingent on the absence of oxygen or other oxidizing species.

Copper (C11000) and Brasses Copper faired poorly in tests, with severe corrosion in the vapor phase and interface of specimens partially immersed in 10–60% HF purged with nitrogen containing 1% oxygen.[22] Rates in the liquid phase were lower but higher than those of Alloys 400 and 715 (C71500). Copper and its alloys (except for the cupronickels) suffer from velocity effects, dealloying, or season cracking.[23] Copper and brasses are therefore of little use in handling HF in liners or processing equipment.

Copper-Nickel Alloys (C70500 and C71500) 90-10 cupronickel (C70500) and 70-30 cupronickel (C71500) have useful corrosion resistance to aqueous HF *provided* the acid is free of oxygen and/or oxidizing agents. Corrosion is highest at concentrations over 10% and less than 70%, peaking between 20 to 40%. Figure 4.29 shows graphically the results of laboratory corrosion tests of Alloys 400, 715, and copper exposed to concentrations of up to 60% HF at 60°C (140°F) under an air blanket and with no agitation. Corrosion rates were lowest in acid below 10% and at 60%. At 60°C (140°F), corrosion rates in the vapor phase and at the liquid-vapor interface are likely to be considerably higher unless the oxygen content in the vapor is less than 1% in this concentration range.[22] In tests conducted in boiling 38% and 48% HF, test results showed that copper and the copper-nickels had lower liquid phase corrosion rates

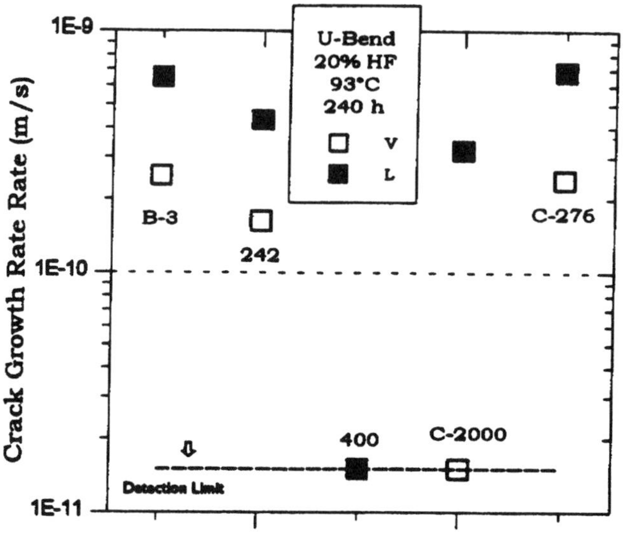

Figure 4.28 SCC Susceptibility of Nickel Alloys in 20% HF at 93°C. For Alloy 400, the Symbols for the L and V Phases Overlap (no detected cracking).

than Alloy 400 (N04400), even with the presence of oxygen in the vapor phase. Copper and the copper-nickel alloys 715 (C71500) and 705 (C70500) also had lower vapor phase corrosion rates until the oxygen concentration in the vapor phase exceeded 2000 ppm (balance nitrogen), at which point the corrosion rates of copper and the copper-nickel alloys were higher than those of Alloy 400.

Interestingly, Alloy 705 (C70500) has been used for piping and equipment in hot (>100°C) 4% HF–2% H_2SO_4 oxygen-free organic streams where it is appreciably more resistant than Alloys 715 (C71500) and 400.

Alloy 715 (C71500) has been used successfully for heat exchanger tubes in AHF coolers that are occasionally washed with water. Although intergranular cracking has been reported at the cold-worked expanded tube ends after many years of service, the alloy is considerably more resistant to SCC than is Alloy 400.

Other Metals

The corrosion behavior of some other metals in hydrofluoric acid is of some interest, despite the prevailing use of nickel-based alloys.

Chemical Lead (L51120) Chemical lead is resistant to *unaerated* aqueous HF solutions below 65% at temperatures up to about 70°C (158°F).[24] It is resistant to 20% HF at the boiling point. HF at a concentration of 40% has been stored in lead containers.[25] Its corrosion rate is considerably increased by a high degree of aeration, particularly in the more dilute solutions, and by velocity. It is also reported in the latter reference

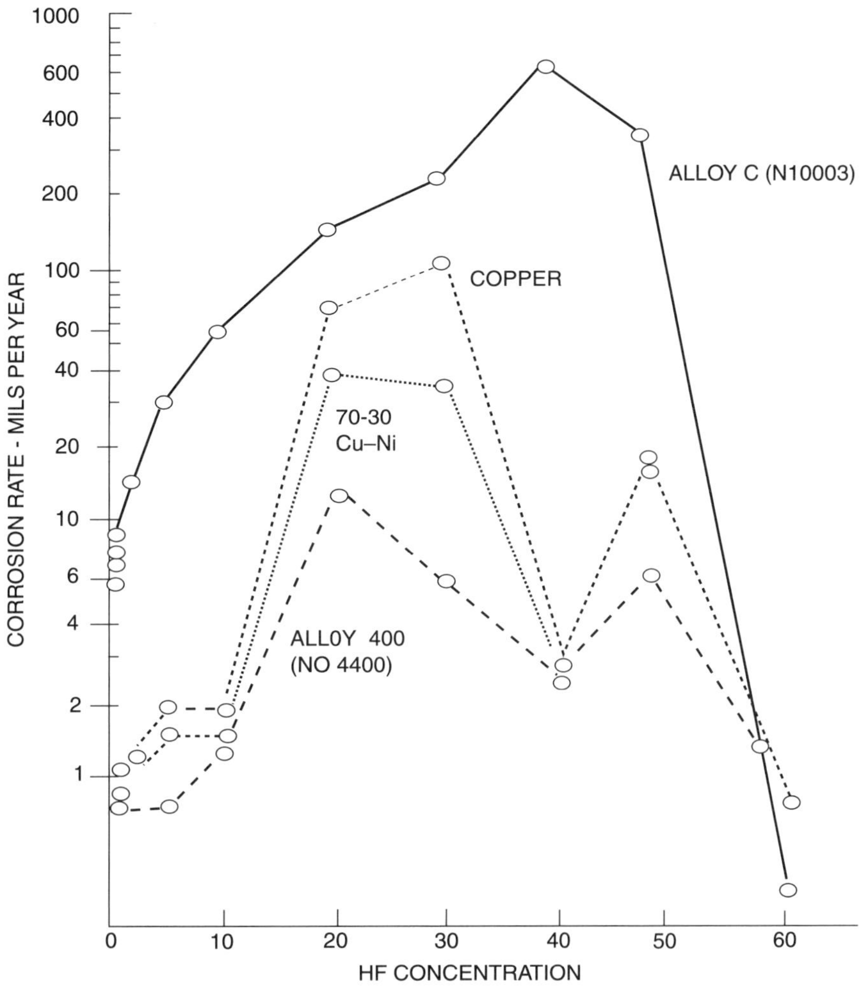

Figure 4.29 Corrosion of Nickel-Copper Alloys in Aqueous HF under Air Blanket at 60°C (no agitation)

that alloying with antimony (L52545) increases the resistance of lead to hydrofluoric acid. Concentrated HF above 65% attacks lead, with rapid attack by AHF. Lead is seldom used anymore and may be considered obsolete because of the health hazard involved in lead "burning" (welding).

Molybdenum (R03600) Molybdenum has good corrosion and erosion-corrosion resistance to aqueous (as well as anhydrous) HF. Corrosion rates in aerated 25–50% HF at 100°C (212°F) are reported to be 16–20 mpy.[26] Rates in the absence of air or oxidizing agents are small to negligible. Although molybdenum and its alloys are difficult to fabricate by welding, molybdenum has been used for valve trim and

Table 4.11 Corrosion of Molybdenum in HF (average of five 48-hr periods)

% HF	Temperature	Aeration	Corrosion (mpy)
25	25°C (77°F)	No	0.1
		Yes	0.2
	100°C (212°F)	No	4.1
		Yes	20.0
49	25°C (77°F)	No	0.1
		Yes	0.1
	100°C (212°F)	No	2.3
		Yes	16.0
		Yes	20.0
49	25°C (77°F)	No	0.1
		Yes	0.1
	100°C (212°F)	No	2.3
		Yes	16.0

parts. Table 4.11 shows the corrosion rate of molybdenum at 25% and 49% HF at 25°C and 100°C.

Tungsten has comparable corrosion resistance to HF and AHF but is most difficult to weld and is brittle at room temperature, except in thin sections, so it is of little practical value.

Silver and Sterling Silver Pure silver (P07020) has excellent resistance to hydrofluoric acid in the presence or absence of oxygen, although a measurable increase in the rate of corrosion was noted when oxygen was added.[22] Corrosion tests conducted in the liquid and vapor phases of boiling 38% and 48% HF and in 50, 65, and 70% HF at 60°C (140°F) F), with the vapor phase purged with 1% oxygen/99% nitrogen showed negligible rates. Sterling silver (P07931; 92.5% Ag–7.5% Cu) had higher but useful rates. For example, the corrosion rates of pure silver in boiling 38% and 48% acid under the above conditions were 0.1–0.6 mpy, while those of sterling silver were 2.5–4.6 mpy. In an HF alkylation plant, silver-clad equipmemt was used for construction of acid regenerators based on laboratory corrosion tests, which showed that silver had the lowest corrosion rate in the constant boiling mixture of HF and water (38% HF). Unfortunately, the traces of sulfur present in the plant acid converted the cladding to silver sulfide, which was attacked by the HF exposing the substrate steel to attack.[7]

Other Precious Metals Precious metals, including gold (P00020) and platinum (P04995), are essentially unaffected by aqueous HF to the boiling point and probably at higher temperatures, although there is limited data to support this assumption.[27]

Magnesium Magnesium has very limited resistance to aqueous HF, although it has useful resistance to concentrations up to 40% at 25°C (77°F). Corrosion rates of 16–25 mpy in 40–50% HF at 60°C (140°F) for a 6% Al–1% Zn (M11610) alloy have been reported but noting that IGSCC is likely to occur with a 5% aluminum alloy.[25]

Nonmetallic Materials

Plastics

Various types of thermoplastics have some application in aqueous hydrofluoric acid even though only the fluorinated variety will resist AHF.

Polyethylene There are three basic types of polyethylene. There is the low-density type made by a high-pressure process, a product of medium density (MDPE), and a high-density material (HDPE), both made by a low-pressure process. Most MDPE and HDPE grades are copolymers incorporating a few percent of butane or hexane.

The copolymer modification of polyethylene results in improvements in creep resistance and environmental stress cracking (ESC) resistance of the product.

The chemical resistance of the medium- and high-density types, which have largely displaced the low density material, is reported to be equal to or better than that of the low-density type.

Polyethylene bottles in wooden cases and steel drums with polyethylene liners are permitted by DOT Hazardous Material Regulations (para.174.264) for shipments of hydrofluoric acid not exceeding 72% strength. However, most grades of polyethylene are susceptible to ESC when stressed in certain environments particularly wetting agents, oils, and detergents.

There has been one reported case of ESC of a 400 gallon (152,000-L) HDPE tank used to store 70% HF at ambient temperature.[28] The tank was fabricated in one-piece seamless construction by rotational molding. After two years service, the tank leaked. Stresses were residual from the molding process and were also caused by filling and emptying the tank. Ten strips were cut from the tank and tested in 70% HF according to ASTM D 1693, of which nine failed by ESC.

However, one authority in reply stated that properly cured, cross-linked HDPE tanks in sizes to 12,000 gallons had been used to store 70% HF and have been in service for as long as ten years.[29] In view of the hazardous nature of HF, it is questionable whether a free standing polyethylene tank represents good engineering judgment.

Polyethylene has been used to ship 49% electronic grade HF in gallon bottles and as tank liners in the etching process for production of silicon chips, but PE is now being replaced by fluorocarbon plastics, particularly PFA and PVDF.

Polypropylene Two types of polypropylene are available. One is the homopolymer of propylene and the other is a copolymer with some ethylene. The copolymer has better low-temperature properties, is more resistant to ESC, and is the grade generally used in lined piping. Polypropylene is available as sheet and tubing and also as glass cloth-backed sheet. The sheet can be formed to shapes for vessel or pipe linings, which are then cemented to the steel shell, butt welded and spark tested, using standard methods. Alternatively, the lining can be externally reinforced with FRP to form a column or vessel. Also, vessels and columns can be made by forming and welding (on suitable supports), spark testing the seams and reinforcing with a suitable fiberglass-resin (FRP) reinforcement. However, in view of the hazardous nature of HF, lined steel construction is to be preferred.

Polypropylene-lined steel piping is recommended by the manufacturer for use in handling hydrofluoric acid at concentrations of 60% or less up to a temperature of 93°C (200°F).[30]

Polypropylene-lined piping and equipment should not be used where temperatures may approach 18°C (0°F) because the combination of increasing brittleness and thermal shrinkage of the plastic lining can result in cracking of the lining, generally around the flared end of pipe sections.

Fittings should be lined ductile cast iron or cast steel. Gray cast iron should be avoided because of the brittle nature of the iron casting and the hazardous nature of HF.

There is little, if any, published experience in using polypropylene lined vessels in HF service.

Fluorocarbons The fluorocarbon plastics are suitable for HF service as gaskets and as linings for pumps, valves, piping, and equipment over the entire range of concentrations. Limiting temperatures vary with the specific plastic and end use. The more important fluoropolymers used in the chemical process industry are listed below:

Polytetrafluoroethylene	PTFE
Fluoronated ethylene-propylene	FEP
Perfluoroalkoxy	PFA
Polychlorotrifluoroethylene	PCTFE
Polyvinylidene fluoride	PVDF
Polyvinyl fluoride	PVF
Ethylene tetrafluoroethylene	ETFE
Ethylene chlorotrifluorethylene	ECTFE

PTFE is widely used as lined piping, pumps, valves, bellows, and gaskets. Although not melt-processible, a process has been developed to "forge" butt welds. Cylinders so formed have been used to line flanged carbon steel vessel/column sections.

The melt-processible polymers (FEP, PFA, and PVDF) are widely used in valves and pump linings. FEP and PVDF are also widely used as lined carbon steel piping. FEP and PFA, the latter graphite-filled for improved heat transfer, are available as heat exchangers.

FEP, PFA, PVDF, and ECTFE are used as bonded sheet linings for steel vessels. As in the case of polypropylene, glass cloth is mechanically embedded into one surface. In the case of PVDF and ECTFE, polyester cloth can be used instead of glass. Also, a copolymer with greater flexibility, but reportedly equal chemical resistance, is used for lining. The sheet linings can be cemented to clean, grit-blasted steel surfaces with epoxy resin or neoprene-rubber adhesive and the seams butt welded with hot gas and spark tested.

Cemented sheet-lined tanks and truck trailers lined with PFA and PVDF have been used in aqueous 49% and 70% HF service. ECTFE has been used to loose line a 70% HF storage tank and can be used as a cemented lining, in which case a polyester cloth backing is preferred.

Steel vessels and equipment can also be lined with FEP, ECTFE, PVF, PVDF, and PFA by spraying a dispersion coating, followed by baking in an oven and repeating the process to build up a typical thickness of some 40 mils as a corrosion barrier coating. Carbon cloth has been used to reinforce PVDF linings.

Another type of bonded lining uses PFA polymer, with PTFE as a cheaper diluent and viscosity-modifying filler, applied to a steel substrate over a suitable primer/bond coat. Multiple coats are applied. The applicator has also developed a lining system for severe services at elevated temperatures using a stainless steel mesh welded to the steel shell to which the bonded lining is applied in thicknesses of 60 and 90 mils.

With the strength of the chemical bond and the reinforcement provided by the spot/seam welded stainless mesh, disbonding stresses are high compared to adhesive bonded glass-backed linings. The cost of such lining exceeds the cost of glass-lined steel (which is not useful in HF) but is reported to be less costly than vessels made of "exotic" alloys.

The upper temperature limit for plastic-lined steel equipment is often determined not by the diffusion of HF and water vapor, nor by the chemical and thermal stability of the polymer, but by the complex diametrical and longitudinal stress pattern developed during lining and in service as a result of the tenfold difference in coefficient of thermal expansion between the plastic lining and the shell.

The practical result is that the upper limiting temperature of operation is not that of the material itself. For loose-lined fluorocarbon-lined piping, the suggested upper limits are as follows:

PTFE, PFA	150°C (300°F)
FEP	100–130°C (212–266°F)
PVDF	100–110°C (212–230°F)

The thicker the liner, the higher the limit. The maximum recommended temperature for HF over 48% is 93°C (200°F).

For glass-backed or polyester-backed resins used in lining carbon steel vessels, the suggested limit is 60°C (140°F) for FEP, PFA, ECTFE, and PVDF. In concentrations above 48%, however, this temperature should be lowered to 50°C (122°F) because the high concentration of HF in the vapor phase may permeate the resin and attack the glass backing.

Loose, butt welded PTFE sheet lining can be used at temperatures of at least 120°C (250°F) behind carbon brick or with machined graphite cylinders fitted to restrain diametrical movement of the fluorocarbon lining on cooling. The latter observation is based on successful operation of an HCl stripping column.

Bonded and baked linings have been used for coating agitators, reactors, baffles, and other parts that have been placed in services containing HF. Although laboratory tests on panels coated on one side indicate satisfactory resistance to up to 50% HF at 120°C (250°F), such tests do not take into account the factors discussed above and such linings cannot be recommended at temperatures exceeding about 80–90°C (175–195°F).

An exception is the PFA/PTFE lining applied over a stainless steel mesh. This lining has been reportedly tested in 140°C (285°F) concentrated HCl under pressure (with water running over the uncoated external surface) for six months, with no evident penetration indicated upon destructive examination.

PTFE, PFA, FEP, and ECTFE are chemically resistant to all concentrations of HF to their normal maximum use temperatures. PVF and PVDF are limited to 120°C (250°F) in dilute HF (37% or less) and 93°C (200°F) for higher concentrations up to AHF. ECTFE is chemically resistant to 140°C (285°F) in dilute HF and to 120°C (250°F) in 30–49% HF.[31] Although not tested at higher concentrations, it has been used for storing 70% acid. ETFE is resistant to 35% HF to 135°C (275°F), 70% to 120°C (250°F), and AHF to 110°C (230°F).[32]

Fiberglass-Reinforced Polyester Plastics

Fiberglass-reinforced plastics (FRP) are seldom used in hydrofluoric acid service. Glass reinforcement may be attacked by HF as a result of permeation. All resins are attacked at concentrations over 40% at ambient temperatures and an all-plastic structure should not normally be used to contain such a hazardous chemical.[33] However, there are exceptions. A column constructed of a vinyl ester with carbon surface veils has been used in a service containing 30% HF at 55°C (130°F). Storage tanks have been constructed with UPVC (unplasticized polyvinyl chloride) reinforced with FRP. One of the problems is keeping the thin, brittle corrosion barrier on the FRP working surface intact. The UPVC liner (0.125–0.160″ thick) is formed, welded, and spark tested before FRP reinforcement. Liners of polypropylene and various fluorocarbon plastics can also be used, as described in the above sections. UPVC is reported to be resistant to about 50% HF.

All FRP equipment must be designed and built to pertinent standards, such as the "ASME Fiberglass Reinforced Thermoset Plastic Corrosion Resistant Equipment Standard" and the "Canadian Standard for FRP Equipment 41GP22." External loading, such as from attached piping, must be minimized. Also, the safety aspects and the consequences of leaks and structural failure must be carefully evaluated for each application. Filament-wound columns and tanks, for example, must not be suspended by supports attached to the sides of the equipment.

Carbon-Filled Phenolic Plastic Vessels and columns are available constructed of phenolic or furan resin reinforced with short-staple ceramic fiber or carbon filler. For services containing HF, only the carbon-filled phenolic plastic is recommended. Equipment, generally tanks or columns, is made by mixing the resin with coke flour and formed into shape by compacting onto the inside of steel shells. Items are cured, while still inside the steel shell, in the temperature range of 180–290°C (355–555°F) for 24 hours. Equipment is armored by filament winding, using glass-reinforced epoxy resin.

Piping is also available but is a questionable choice for aqueous HF, in view of the brittleness and the high coefficient of thermal expansion of FRP.

The carbon-filled phenolic is rated as excellent in 38% and 50% HF at 110°C (230°F) and good in 70% HF at 60°C (140°F).

In view of its brittleness, all of the design and installation requirements recommended for FRP should be observed as a minimum, and particular consideration should be given to safety aspects.

Rubber and Elastomers

Both natural rubbers and synthetic elastomers are used, within specific limitations, in hydrofluoric acid. Such materials must be specially compounded for this service. Note also that steel vessels to be lined must be designed and fabricated to this purpose (see Chapter 7).

Soft and Hard Natural Rubbers Soft natural rubber has been used for resisting aqueous HF concentrations of 49% and less. A soft natural rubber lining was used in tank trucks shipping electronic grade 49% acid, since it caused less contamination than chlorobutyl.

Information from the literature and suppliers' chemical resistance tables on suitability of natural rubber for various concentrations of HF is fragmentary at best. One recent supplier's chemical resistance table shows semi-hard natural rubber to be fair in hydrofluoric acid and flexible ebonite (hard) to be good. NACE Technical Committee Report 61-13 "Rubber Linings, Hard and Soft for Resistance to Corrosion,"[34] shows soft natural rubber to have a limited recommendation rating in hydrofluoric acid and hard rubber to be recommended. The maximum permissible temperature is shown to be 66°C (150°F). The concentration of hydrofluoric acid may be assumed to be 48–49%, a common laboratory reagent grade.

Butyl, Chlorobutyl, and Bromobutyl Linings This family of elastomeric linings has excellent resistance to aqueous HF. Chlorobutyl is the elastomer of choice for lining tank cars, storage tanks, and tank trucks for use in 70% HF, at least in the United States. However, the chemical resistance of the three types is much the same. Bromobutyl is reported to have better lining and curing characteristics, initial tack and better tolerance to water in curing. Table 4.12 compares the property changes in various elastomers exposed in 50% HF at 22°C (72°F) for 30 days.[35]

Although chlorobutyl elastomer has good resistance to 70% HF liquid, its resistance in the vapor phase is only fair, and the life expectancy of linings is only about five years. The reason for this short life expectancy is that the concentration of HF in the vapor phase above 70% HF is about 98% HF, as can be seen from the graph of Figure 4.30[36] The vapor phase temperature should not exceed 32°C (90°F). Tanks, tank cars, and tank trucks should be painted white to keep the temperature of vessels exposed to sunlight as low as possible.

As the lining deteriorates, particles contaminate the acid. To minimize the prob-

Table 4.12 Property Changes in Elastomers in 50% HF at 22°C for 30 Days

Elastomer	% Weight Change	% Volume Change
Chlorobutyl	5.24	6.02
Chlorobutyl/natural blend	2.57	2.57
Bromobutyl	2.03	2.02
Chlorinated Polyethylene	40.86	38.19

lem, a white filler compound is sometimes used. Since many white pigments react with HF, only linings specifically compounded for HF service should be used.

Some formulations of these butyl rubbers cannot be spark tested due to their carbon content. It is important that the particular formulation from the specific contractor be checked on this point.

In recent years, tank trucks are being lined with PVDC or PFA glass-backed sheet linings for both 49% and 70% concentrations where preservation of product quality is important.

For dilute concentrations of hydrofluoric acid (below 50%), properly compounded elastomers of this type should be satisfactory to temperatures of about 70°C (158°F), although there is a lack of data on this point. It is advisable to conduct confirmatory tests on the selected compound.

Other Elastomer Linings

Chlorosulfonated polyethylene has excellent resistance to hydrofluoric acid of all concentrations. Manufacturer's literature indicates that AHF has "little or no effect" on the elastomer at room temperature.[37] CSPE-lined hose is used for unloading 70% HF tank trailers. Its lack of popularity as a liner may be because it is more difficult to obtain a satisfactory lining than with chlorobutyl and that the lining stock is compounded for other services (e.g., nitric, chromic, and sulfuric acids).

Chloroprene is reportedly suitable for exposure to 48% HF at ambient temperature. However, there appears to be little field experience and confirmatory tests should be run.

Ethylene-propylene (EP) and ethylene-propylene diene monomer (EPDM) are elastomers based on ethylene, propylene, and a controlled amount of non-conjugated diene. They are resistant up to 48% HF at room temperature, the acid having only a minor to moderate effect. EPDM-lined flanged steel ditch sections have been used successfully in an HF manufacturing plant to protect deteriorating brick lining. EPDM elastomers have excellent resistance to sunlight and weathering.

Fluoro- and Perfluoroelastomers Fluoroelastomers have excellent resistance to all concentrations of aqueous HF and can be used for 0-rings and other components. However, it is important to select the proper compound and curing agent. For example, curing with lead oxide (PbO) is to be preferred to magnesium oxide (MgO),

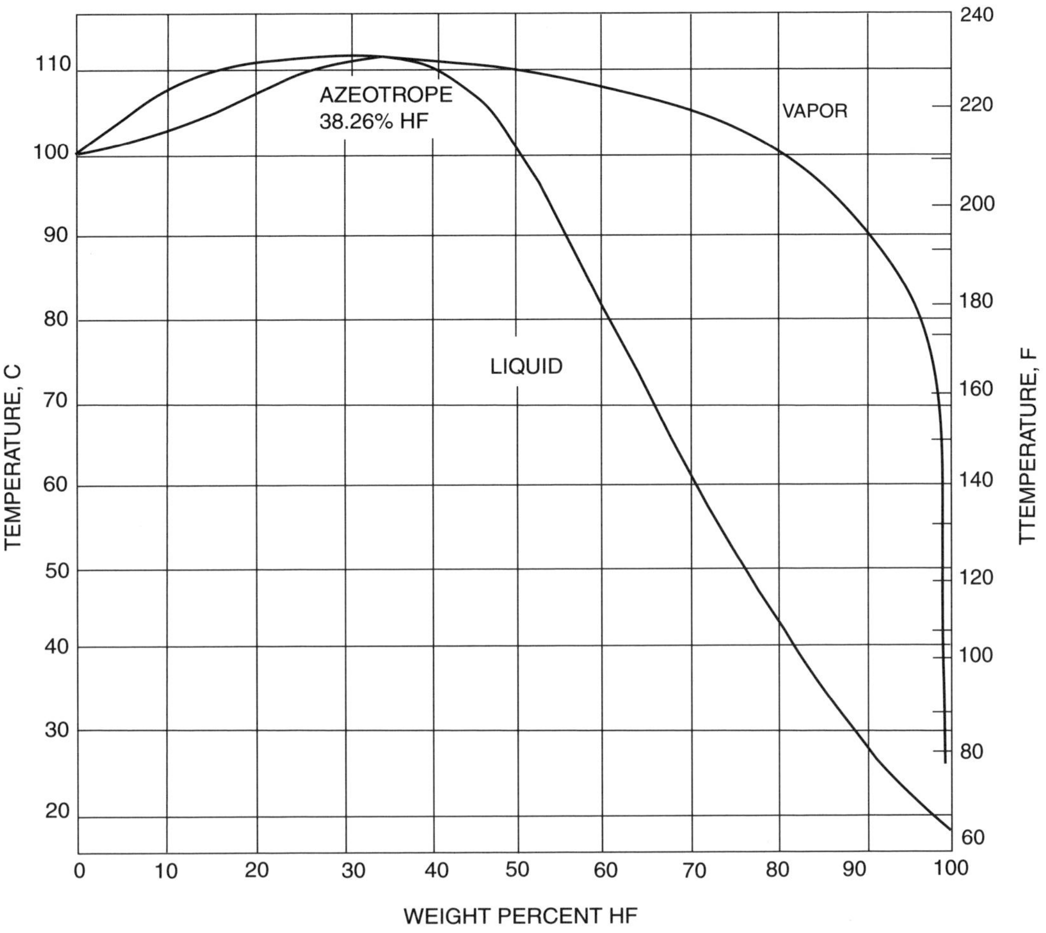

Figure 4.30 Boiling Points and Liquid Vapor Compositions of the System HF-H_2O

which is used for high temperature applications but which has seriously adverse effects on acid resistance. The manufacturer should be consulted to select the best compound and curing agent for the service.

Exposure data for Du Pont's fluoroelastomer in 48% and 75% HF are shown in Table 4.13.[38]

The perfluoroelastomer is more resistant, with types that are little affected (less than 10% swell), and suffers no significant loss of properties affecting performance in 75% HF to temperatures above 100°C (212°F).[39]

Summary The limiting temperatures and concentrations for nonmetallic materials of construction are shown in Figure 4.31.[21]

Table 4.13 Fluoroelastomers vs. Hydrofluoric Acid

% HF	Temp°C (°F)	Days	Type	ΔHardness*	ΔVolume %
48	20° (68°)	7	A–B	0 to −1	0 to 0.7
		21	A–B	−2 to −4	2
75	70° (158°)	5	A	−9	
	100° (212°)	5	A	−13	

*Durometer A Points
A–66% Fluorine (Dipolymers of Hexa-fluoropropylene and PVDF)
B–68% Fluorine (Terpolymers of HFP, PVDF, and PTFE)

Carbon and Graphite

As in the case of AHF, high-purity carbon is suitable for pump seal parts and other non-pressure containing applications, and flexible-graphite gaskets can be used for aqueous grades of HF.

In addition, carbon-brick linings can be used for tanks, columns, ducts, and pipelines in environments containing aqueous HF at temperatures and concentrations limited by the chemical and temperature resistance of the mortar and membrane and the design constraints required to keep the lining in compression.

The carbon brick should also be of low ash content; 1% maximum is preferred but is not always available.

Carbon-brick linings must always be designed with a corrosion-resistant membrane between the lining and the steel. This membrane must be completely resistant to the chemical service. The brick lining provides protection against erosion and mechanical damage and some thermal protection, although carbon brick has higher thermal conductivity than acid brick. Carbon brick is also comparatively porous as well as expensive.

Membranes for hydrofluoric acid service are often chlorobutyl elastomer or PTFE (welded).

Graphite, which is made by heating formed carbon shapes to temperatures in excess of 2500°C (4530°F), has good resistance to aqueous HF. The graphite product is porous and must be impregnated with a suitable resin to make it relatively impervious and useful for chemical plant equipment. This may be accomplished by vacuum treatment followed by impregnation with a phenolic or furane resin under pressure. Another method uses impregnation with PTFE. For high-temperature services, porosity can be reduced by impregnating with pitch followed by pyrolysis. This product is *only* for high temperature applications where complete imperviousness is not required.

Equipment made with "impervious" graphite includes heat exchangers, columns, absorbers, rupture discs, pumps, and other parts.

Graphite is a weak and brittle material. Heat exchangers made from cylindrical or cubical blocks are safer and are preferred for HF service because they can be totally enclosed in a steel shell with lined steel heads. A disadvantage of the block type units is that the mechanical working of the petroleum coke/pitch mass prior to

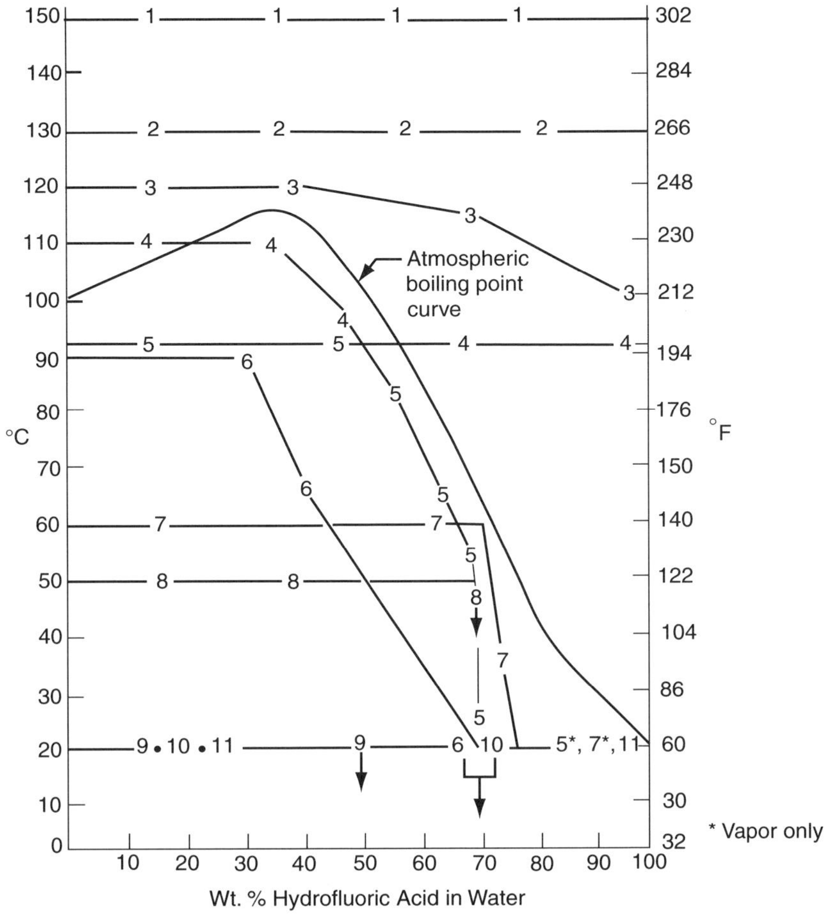

Figure 4.31 Plastics and Elastomers in Hydrofluoric Acid

pyrolysis tends to be more variable when forming blocks than when forming tubular shapes. Consequently, the porosity to be sealed tends to be larger.

Impervious graphite has long been used in HF service, despite potential hazards from the possibility of breakage.

Graphite is immune to thermal shock, but susceptible to failure caused by steam hammer.

PTFE-impregnated graphite has appreciably better chemical resistance to hydrofluoric acid than the phenolic-impregnated grade and to much higher temperatures. However, unlike the phenolic impregnated grade, the PTFE-impregnated material has measurable gas permeability. Therefore, the proposed installation should be discussed with the manufacturer and the tolerable leakage assessed. The concentration and temperature limitations given by the manufacturers are as follows:

- Phenolic-Impregnated Graphite:
  48% HF to its atmospheric boiling point; maximum wall temperature 160°C (320°F)
  60% HF to 85°C (185°F)
- PTFE-Impregnated Graphite
  80% HF; maximum wall temperature 230°C (445°F)

Acid-Resistant Mortars

The preferred mortar is a silica-free, carbon-filled phenolic. This mortar should be resistant to up to 49% HF at the atmospheric boiling point and to 50-70% at a maximum temperature of 110°C (230°F). Furan mortars are less resistant. Silica-free sulfur mortars are also suitable up to 70 % HF at a maximum temperature of 90°C (195°F). One should consult with the selected supplier on the limitations of his products.

References

1. "Hazardous Materials Regulations Pertaining to Hydrofluoric Acid and Hydrogen Fluoride," (49CFR), Parts 178–179. Department of Transportation (DOT).
2. G.C. Whitaker, *Corrosion*, Vol. 6, No. 9, p. 283 (1950).
3. W.K. Blanchard and N.C. Mack, Paper 452, CORROSION 92, NACE, Houston, TX.
4. International Nickel Co. Inc. "Corrosion Engineering Bulletin CEB-5" 1975.
5. J. Fritz, Allegeheny Ludlum, Private Communications, 2001.
6. R.B. Rebak, Presentation to NACE International Committee STG 36, CORROSION 98.
7. R.B. Rebak, J.R. Dillman, P. Crook and C. Shawber, "Corrosion Behavior of Nickel Alloys in Wet Hydrofluoric Acid," ACEMA 2000.
8. S.J. Pawel, *Corrosion*, Vol. 50, No. 12, December 1994, p. 963–971.
9. T.C. Spence, "Cast Ni-Cu Alloys in HF," presentation to NACE International Committee STG 36 in Corrosion/97, March 11, 1997.
10. H.S. Jennings, CORROSION 2001, Paper 01345, NACE, Houston, TX.
11. W.K. Blanchard and N.C. Mack, Paper 452, CORROSION 92, NACE, Houston, TX.
12. N. Sridhar, et al., *The Arabian Journal for Science and Engineering*, Vol. 14, No. 2 (April 1989).
13. N. Sridhar, Paper No. 19, CORROSION 90, NACE, Houston TX.
14. J.R. Crum, G.D. Smith, M.J. McNallan, and S.Hirnyl, Paper 382, CORROSION 99, NACE , Houston, TX.
15. G. Schmitt, S.Losaker, "Performance of CRA in Hydrofluoric Acid," EUROCORR '97 Proceedings, Volume 1, European Federation of Corrosion, Trondheim, Norway, p. 677–682.
16. B.E. Paige, *Materials Performance*, Vol. 15, No. 12, p. 22 (December 1976).
17. T.F. Degnan. *Process Industry Corrosion—The Theory and Practice*, National Association of Corrosion Engineers, p. 285 (1986).

18. M. Schussler, "Metal Materials for Handling Aqueous Hydrofluoric Acid," *Ind. Eng. Chem.*, 47,135, 1955.
19. H.R. Copson and C.F. Cheng, *Corrosion*, Vol. 12, No. 12, p. 647–653, 1956.
20. L. Graf and W. Wittich, *Werkstoffe und Korrosion*, Vol. 17, p. 385 (1966).
21. NACE International Publication 5A171 (latest Revision), "Materials for Receiving, Handling, and Storing Hydrofluoric Acid," Houston, NACE International.
22. H. A. Pray et al., Battelle Memorial Institute Report BMI-268 (Del.), "Corrosion-Resistant Materials for Hydrofluoric Acid," (15 June 1953).
23. Philips Petroleum Company, "Hydrofluoric Acid Alkylation," 1946.
24. W.Z. Friend and H.O. Teeple, *Oil and Gas Journal*, 16 March 1946.
25. E. Lingnau, *Werkestoff u. Korrosion*, No. 4, p. 216, 1957.
26. Climax Molybdenum Co., "Molybdenum Metal," Technical Notes, November 1959.
27. H.H. Uhlig, *Corrosion Handbook*, John Wiley and Sons, p. 316 (1948).
28. J.H. Van Sciver, *Materials Performance*, Vol. 29, No. 12, p. 49. (December 1988).
29. G.E. Carrow, *Materials Performance*, Vol. 2, No. 5, p. 74. (May 1989).
30. Dow Chemical Company Bulletin, "Chemical Resistance Guide for Systems Using Dow Plastic Lined Piping" (1990).
31. Ausimont, USA Inc. "Chemical Resistance of Halar7 Fluoro-carbon."
32. E.I. DuPont DeNemours & Co., "Chemical Use Temperature Guide, Tefzel7 Fluoropolymer."
33. Ashland Chemical Co., "Hetron and Aropol Resin Selection Guide for Corrosion Resistant RTP Applications," Bulletin 1545.
34. NACE Technical Report 61.13, "Rubber Linings, Hard and Soft for Resistance to Corrosion," NACE, Houston, TX.
35. L. DeLashmit, "Rubber Linings for Hydrofluoric Acid," presentation to NACE International Committee T5A-36, CORROSION 2000, NACE, Houston, TX.
36. E.I. DuPont de Nemours & Co., Bulletin "Hydrofluoric Acid, Anhydrous Technical Properties, Uses, Storage and Handling."
37. E.I. DuPont de Nemours & Co., "Chemical Resistance Chart Hytrel, Neoprene, Hypalon, Nordel, Viton, Vamac."
38. E.I. DuPont de Nemours & Co., "Viton Fluoroelastomer Resistance Guide," VT-500.1 (9/90)K.
39. E.I. DuPont de Nemours & Co., "Kalrez! Perfluorocarbon Chemical Resistance and Fluid Compatibility" 10/1/88.

5

High Temperature Corrosion Behavior in HF and AHF

Processes such as the production of fluoropolymers and fluorochemicals and the destruction of fluorocarbon waste involve the exposure of materials of construction to hydrogen fluoride (HF) at elevated temperatures. Containing this highly corrosive and highly toxic gas presents a significant challenge to the materials engineer.

Metals and Alloys

The data presented in this section should be used as a guide for longer-term investigations at high temperature. Most of the published data for metals is from exposure tests of a few hours to a few days, which is appropriate for ranking relative corrosion resistance and observations of short-term physical changes in the metal. Longer-term testing above 400°C can lead to metallurgical changes in some alloys that affect the mechanical and corrosion properties.

Corrosion-Resistant Alloys

In the presence of HF, some metals will form a metal fluoride film that may be thin and adherent and provide protection to the metal.[1] However, at temperatures where the fluoride is molten, a high corrosion rate will result due to the dissolution of metal and oxide in molten fluoride. For fluorides with high vapor pressures, the corrosion rate will be strongly influenced by the velocity of the corroding gas due to mass transfer effects. Another feature of the interaction of HF with materials is the high permeability (solubility-diffusivity product, $N_F D_F$) of F in metals. For example, the permeability of fluorine in nickel is approximately five times that for oxygen in nickel and about two times that for chlorine in nickel at 750°C. This means that metals will readily form internal metal fluorides in HF-containing environments. This can result in a degradation in the mechanical properties of the materials and in deterioration in its resistance to other corrodents.[1]

Myers and DeLong[2] conducted tests on nickel, nickel-based alloys, and stainless steels in AHF at 500°C and 600°C. The data in Table 5.1 show very high corrosion rates on stainless steels relative to nickel and nickel-based alloys. The stainless steels with higher chromium and niobium (309, 310, and 347) showed the highest corrosion rate at 600°C. Myers and DeLong also exposed various metals in 50% HF and 50% steam, from 550 to 750°C. The data in Table 5.2 show increasing corrosion of both Alloy 200 and Alloy 400 with temperature. At 700° and 750°C Alloy 400 had higher corrosion rates than Alloy 200. In comparison with data in Table 5.1, the data indicate that steam in HF increases the corrosion rate of Nickel 200, however, it has an inhibiting effect on Alloy 400.

Data from longer-term tests conducted at 500–600°C in HF on a range of commercial nickel-based alloys are presented in Table 5.3[3]. The longer test period data show lower corrosion rates than those in Table 5.2. Alloy C exhibited the lowest corrosion rate, followed by Alloy 600 and Alloy B. Crum, et. al.[4] reported on the behavior of alloys in an HF gas stream at 450°C. The composition of the gas stream was reported to be 5% HF, 8% O_2, 15% H_2O (balance CO_2 and N_2). The corrosion rates presented in Table 5.4 are very low for all alloys. This is thought to be due to

Table 5.1 Corrosion Rate of Various Metals in AHF at 500°C and 600°C, Tests 4–15 Hours

	Corrosion Rate, mm/y (mpy)	
Material	500°C	600°C
Alloy 200	0.9 (36)	0.9 (36)
Alloy 400	1.2 (48)	1.8 (72)
Alloy 600	1.5 (60)	1.5 (60)
Copper	1.5 (60)	1.2 (48)
Carbon steel (1020)	15.5 (612)	7.62 (300)
SS 304	—	13.4 (528)
SS 347	183 (7200)	177 (6900)
SS 309Cb	5.79 (228)	168 (6600)
SS 310	12.2 (480)	305 (12000)
430	1.5 (60)	11.6 (456)

Table 5.2 Corrosion of Alloy 200 and Alloy 400 in 50% HF and 50% Steam, Tests 4–15 Hours

	Corrosion Rate mm/y (mpy)				
Metal	550°C	600°C	650°C	700°C	750°C
Alloy 200	0.8 (31)	1.8 (72)	2.7 (108)	3.7 (144)	3.1 (120)
Alloy 400	—	0.6 (24)	1.5 (60)	4.0 (156)	5.2 (204)

Table 5.3 Corrosion Tests in Hydrogen Fluoride Gas 500–600°C, 36 Hours*

Alloy	Corrosion Rate, mm/y (mpy)	Comments
Alloy C	0.01 (0.3)	Iridescent tarnish film
Alloy 600	0.02 (0.7)	Iridescent tarnish film
Alloy B	0.05 (2)	Black film
Alloy 200	0.23 (9)	Black film
Alloy 201	0.36 (14)	Black film
Alloy 400	0.33 (13)	Adherent dark film
Alloy K-500	0.41 (16)	Adherent dark film

*7 lb HF at 4 psig was passed through a laboratory furnace for hydrofluorination of metal oxides.

Table 5.4 Corrosion Rate of Various Alloys tested in HF Gas Stream at 450°C

	Corrosion Rate, mm/y (mpy)		SCC Yes/No	
Material	100 hours	155 hours	100 hours	155 hours
Alloy 600	0.01 (0.4)	0	Yes	Yes
Alloy 400	0	0.002 (0.1)	No	No
Alloy 825	0.01 (0.4)	0	No	No
Alloy C-276	0	0	No	No
Alloy 686	0	0	No	No
Alloy 622	0	0	No	No
Alloy 25–6Mo	0	0	No	No

*Same specimens evaluated at 100 hours and after additional 55 hours exposure.

water vapor and oxygen promoting a protective metal oxide on these alloys In addition, Alloy 600 was reported to have undergone stress corrosion cracking.

Several Monel alloys were studied by Tyreman and Marsh[5]. Monel 400 exhibited the formation of manganese and iron difluorides within the alloy matrix. Monel K-500 underwent rapid deterioration in the range 700–850°C due to the preferential formation of aluminum trifluoride from the gamma prime ($Ni_3(Ti,Al)$) phase within the substrate material.

Marsh and Elliott[6] investigated the effect of chromium content on the resistance of Ni-Cr alloys to HF and HF + steam at 650°C and examined the behavior of commercial nickel-based alloys. These authors found that the corrosion rate increased with increasing chromium content of an alloy. In anhydrous HF, chromium fluorides developed beneath thin external corrosion product scales and for the nickel-chromium alloys an increase in the chromium content increased the rate of reaction. In the HF + steam environment the alloys formed either an external Cr_2O_3-rich scale with internal fluoride in the underlying alloy or internal Cr_2O_3-rich areas within a corrosion product network also containing chromium fluorides. The authors cited work by Sturm,[7] which showed that a low melting point eutectic mixture exists at CrF_2 + 14 mol % CrF_3. This eutectic melts at 831°C, much lower than the melting points of pure CrF_2 (894°C) and CrF_3 (1404°C).

One of the few studies to focus on the behavior of materials at temperatures above 800°C was carried out by Tyreman & Elliott[8] who evaluated the behavior of pure metals and alloys in HF and HF-steam mixtures at 850°C. Nickel, cobalt, and copper were the most resistant pure metals, while chromium was heavily attacked and tantalum and niobium were completely destroyed in 10 hour tests. Nickel-based alloys were generally better than Fe-base alloys, which corroded rapidly by internal fluoridation and heavy oxide formation resulting from pyro-hydrolysis of fluorides in steam. Reactive elements such as Nb were deleterious for some Ni-based alloys. Mo was found to be beneficial in the Ni-Cr-Fe system.

In the study by Tyreman & Elliott[8] molybdenum showed negligible attack by HF after 10 hours at 850°C. Molybdenum was also tested by Zotikov and Semenyuk[9] who measured its corrosion rate in HF containing 0.6% moisture as 51 mpy at 800°C. The corrosion rate of tungsten in the same environment increased dramatically above 600°C, from 0.9 mpy at 600°C to 36 mpy at 700°C.

Barnes[1] investigated pure metals, alloys, and ceramics after exposure in Ar-5% HF for 15 hours at 1000°C. The pure metals tested included three grades of nickel (Alloy 200, Alloy 201, and Alloy 270), copper, gold, platinum, and palladium. The corrosion rate results are listed in Table 5.5. Barnes reported that the precious metals and copper had little change in appearance after exposure in HF. The gold and platinum samples underwent etching of the grain boundaries. Palladium had no

Table 5.5 Corrosion Rate and Observations on Pure Metals Exposed in Ar-5% HF 1000°C for 15 Hours

Material	Weight Change (mg/cm^2)	Corrosion Rate (mpy)	Appearance	Comments
Gold	0	0	Unchanged	No evidence of internal fluorides
Platinum	0.2	1.1	Matt Gray	No evidence of internal fluorides
Palladium	0.2	—	Unchanged	No evidence of internal fluorides
Chromium	−35.25	338.2	Green	Extensive internal fluorides
Alloy 200	−0.1	1.4	Unchanged	No evidence of internal fluorides
Alloy 201	−0.1	1.4	Unchanged	No evidence of internal fluorides
Alloy 270	0	0	Unchanged	No evidence of internal fluorides

*Chromium exposed for 50 hours at 1000°C in Ar-5% HF.

Table 5.6 Metallographic Measurements and Observations on Alloys Exposed in Ar-5% HF 1000 Hours for 15 Hours

Alloy	Maximum Depth of Affected Metal, mm/side (mils/side)	Appearance	Comments
242	0.015 (0.6)	Surface scale	Also resistant in Ar-35%
Alloy B-3	0.020 (0.8)	Surface scale	—
IC-221M Nickel Aluminide	0.025 (1.0)	Surface scale	Extensive Corrosion in Ar-35% HF
Alloy 188	0.051 (2.0)	Surface scale	—
Alloy 617	0.13 (5.3)	Nodules on surface	Extensive internal fluorides
Alloy 600	0.15 (5.9)	Nodules on surface	External internal fluorides
Alloy 602 CA	0.18 (7.1)	Surface scale: Molten corrosion product	External internal fluorides

apparent change in this environment but was heavily corroded and formed a molten corrosion product in Ar-35%HF. Copper retained its metallic luster after exposure.

Nickel samples appeared generally unchanged after exposure in HF, retaining their metallic luster. The high purity nickel Alloy 227 showed no degradation. SEM examination of the surface revealed etching of grain boundaries but no evidence of a surface scale. The nickel samples examined after exposure periods up to 50 hours in Ar-5% HF did not exhibit measurable internal fluoride formation. A nickel sample holder that had been exposed for approximately 1600 hours in Ar-5% HF at 1000°C was sectioned and the depth of nickel fluoride particles was found to be approximately 20 μm. The pure chromium sample exhibited a brittle surface scale of chromium fluoride and extensive internal fluoride formation. Internal fluoride extended to a depth of approximately 250 μm after 50 hours at 1000°C in Ar-5% HF. The sample was extremely brittle after exposure.

Barnes[1] reported that nickel- and cobalt-based alloys had higher corrosion rates than pure nickel and exhibited internal fluoride formation. The test data is presented in Table 5.6. There was a substantial difference in the behavior of the alloys tested. Inconel 600 and 617 formed surface nodules of chromium fluoride and also had extensive internal fluoride. HA188, HA 242, and B-3 had thin surface scales and less internal fluoride formation than the other alloys. The 602CA alloy formed a thick fluoride scale together with extensive internal fluoride formation. The intermetallic IC-221M had good resistance at 5% HF but exhibited poor resistance at higher concentrations. In Ar-35% HF the material was almost completely converted to fluoride.

Table 5.7 Weight Change and Observations of Ceramics and Graphite Exposed in Ar-5% HF 1000 hours for 15 hours

	Weight Change (mg/cm^2)	Corrosion Rate (mils/year)	Comments
Al_2O_3 (polycrystalline)	−2.4	141	General wastage
Al2O3 (sapphire)	−0.9	52	Frosted appearance
SiC	−5.7	411	General wastage
Si_3N_4	−2.2	155	General wastage
Graphite (<5 ppm ash)	0	0	Unchanged
Graphite (700 ppm ash)	0	0	Unchanged

Ceramics and Graphite

Barnes[1] studied the behavior of ceramics and several grades of graphite in HF-containing environments at 1000°C. The ceramics tested were sapphire, polycrystalline alumina, sintered silicon carbide, and silicon nitride. The two graphite grades contained <5 ppm and 700 ppm ash, respectively. The behavior of the materials is summarized in Table 5.7. The graphite samples exhibited little change after exposure in HF. The sapphire sample, which was received as a translucent disc, became opaque in HF. Alumina, silicon carbide and silicon nitride all suffered general wastage.

References

1. J.J. Barnes, "Materials Behavior in High Temperature HF Containing Environments," Corrosion/2000, Paper 518 (Houston, TX: NACE International, 2000).
2. W.R. Myers and W.B. DeLong, *Chem. Eng. Prog.*, 44, 359 (1948).
3. "Corrosion Engineering," Bulletin CEB-5, International Nickel Co, Inc., 1975.
4. J.R. Crum, G.D. Smith, M.J. McNallan, and S. Hirnyj, NACE Corrosion 99, Paper #382.
5. G. Marsh, Ph.D. Thesis, University of Manchester, 1982; C.J. Tyreman, Ph.D. Thesis, University of Manchester, 1986.
6. G. Marsh and P. Elliott, High Temperature Corrosion in Energy Systems, M.F. Rothman (ed.), The Metallurgical Society of AIME, Warrendale, PA (1985), 467–481.
7. B.J. Sturm, *Inorganic Chemistry*, 1(3), 665 (1962).
8. C.J. Tyreman & P. Elliott, NACE Corrosion '88, Paper # 135.
9. V.S. Zotikov and E. Ta Semenyuk, Metallschutz, Moscow, Vol. 6, p. 218 (1970).

6

Hydrofluoric Acid: Corrosion in Mixed Acids

Earlier chapters discussed the corrosion of various materials in essentially pure hydrofluoric acid environments. Data in the literature has shown that fluorides have a notable effect when present as a minor constituent in other media. This chapter discusses various acid mixtures.

Corrosion by Mixtures of Sulfuric Acid and HF

Mixtures of sulfuric and hydrofluoric acids are encountered in the manufacture of HF and of some organic fluorides.

Experience has shown that when the total acidity (HF + H_2S0_4) is high, the more corrosion-resistant nickel-based alloys containing chromium, molybdenum, and copper are useful materials of construction. Alloys of this class include Alloy 20, Alloy 825, Alloy G-3, and Alloy G-30.

Shridhar[1] using electrochemical and immersion corrosion test techniques studied the corrosion behavior of various alloys exposed to sulfuric acid with HF additions. The effect of velocity was also examined using rotating cylinder electrodes. A graphical presentation of some of his test data, taken from his paper, is shown in Figure 6.1.

The study made the following conclusions:

- In high-nickel-molybdenum alloys containing chromium, HF additions inhibit corrosion in concentrated sulfuric acid and accelerate corrosion in dilute sulfuric acid. Data generated for Alloy G-30 clearly demonstrates this behavior (see Table 6.1).
- The transition, or crossover, occurs between 60% and 80% sulfuric acid, depending upon the presence of copper in the alloy. The presence of copper lowers the crossover concentration from 80% to 60%.
- Velocity has only a small effect in concentrated sulfuric acid containing HF but a large effect in more dilute acid mixtures.

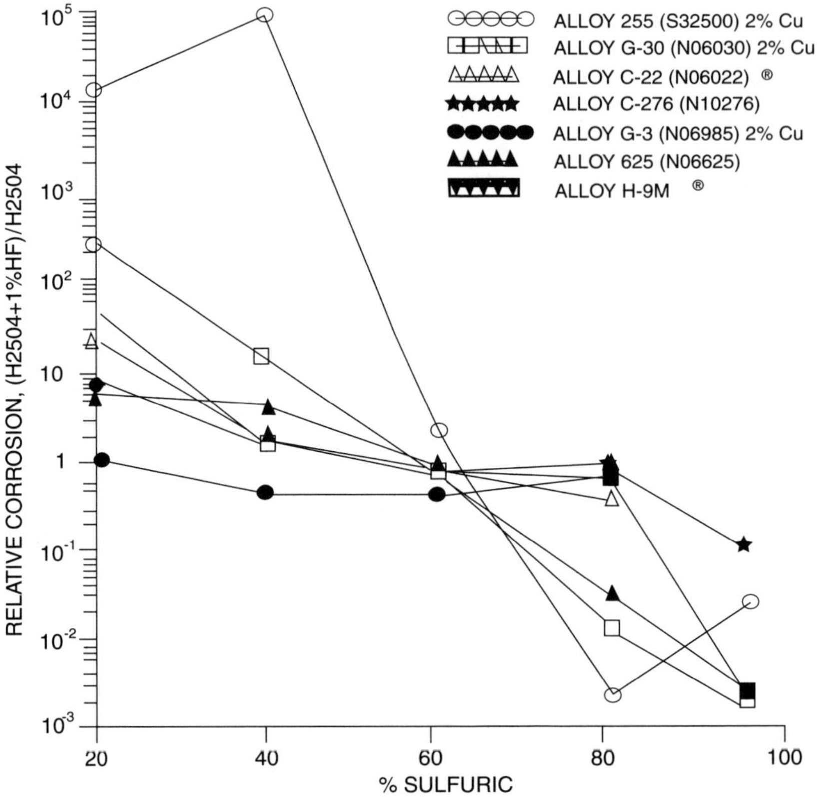

Figure 6.1 Effect of Alloy Content on Switching to Inhibition in H_2SO_4 + 1% HF at 79°C

Table 6.1 Corrosion Test Data in Sulfuric-Hydrofluoric Acid Mixtures. Alloy G-30 (UNS N06030). Flat unwelded coupons, 24 hours, 70°C.

% H_2SO_4	0% HF	1% HF	3% HF	5% HF
20	0.1	20.3	20.7	20.7
40	0.7	8.7	7.0	8.6
60	11.0	8.2	7.8	9.0
80	122	1.9	0.8	2.0
85	—	—	—	0.1
90	125	—	1.9	—
94	—	0.2	—	—
96	102	—	—	—

- The threshold concentration of HF needed for inhibition lies between 100 ppm and 1% for 90% H_2SO_4. No inhibitive action of HF was observed at 10 ppm.

A Russian paper reports results of corrosion tests of a Cr-Ni-Mo steel (0Kh23N28M3D3T; Fe-23 Cr-28 Ni-3 Mo-3 Cu; Ti-stabilized) exposed in various equipment throughout an HF manufacturing plant.[2] Included was a monohydrate column, absorption column, two evaporators, and a rectification column. A wide range of strong acid mixtures was obtained, ranging from 90–95% sulfuric acid to 96–97% HF. Temperatures ranged from 20 to 70°C (68–158°F). Corrosion rates varied from 0.8 to 8.0 mpy. The alloy is analogous to the original Alloy 20, before the nickel content was increased, except that the alloy is stabilized with titanium, rather than niobium.

Alloy 825 has been recommended for HF production plants.[3] Knife-line attack (KLA) was noted in weld heat-affected zones and thus seamless heat exchanger tubes were recommended. Full-finished welded tubes (i.e., cold-drawn a minimum of 35% reduction area before the final anneal) should also be satisfactory.[4]

Corrosion by Mixtures of HF with HCl

Air has a greater influence than 2% HCl in 10% HF for corrosion of Alloy G-30 and Alloy Al-6XN. Corrosion rates for Alloy G-30 and Al6-XN in 10% HF, 2% HCl, and 10% HF with 2% HCl at 66°C with and without solution purging are presented in Table 6.2.[5]

There is negligible effect from the 2% HCl on the corrosion of these alloys. Both alloys have lower corrosion rates in nitrogen purged HF-HCl solutions. The corrosion rates were found to be erratic if no purge is applied. With air purged through the solution the corrosion rates are significantly higher compared to rates from nitrogen purged solutions.

Nitric-HF Mixtures

Although hydrofluoric acid is a minor constituent in nitric-HF mixtures for bright-pickling austenitic stainless steels, it drastically changes the oxidation characteristics.

TABLE 6.2 Corrosion Rate (mpy) of Alloy G-30 and Al-6XN Exposed to 10% HF and 2% HCl at 66°C—With and Without Purging

	2% HCl	10% HF	10% HF + 2% HCl		
Alloy	No Purge	No Purge	Air Purge	No Purge	N_2 Purge
Alloy G-30	0.3	67	10	46–75	12
Alloy Al-6XN	0.4	115	180	95–356	78

Nitric acid is a powerful oxidizing acid inherently non-corrosive to stainless steels up to about 65% to the atmospheric boiling point and to 95% to about 30°C (85°F).[6] However, the corrosive nature changes in mixtures, typically 10–15 HNO_3-1–3 HF. These mixtures are used to remove the black oxide scale from annealed stainless steel products or, as pastes, to remove the heat tint in the HAZ from welding. Even at the moderate temperatures employed, usually not more than 70°C (150°F), 18–8 stainless steels suffer active corrosion which is aggravated several-fold if the metal is sensitized, due to interganular corrosion (IGA). This is the basis for Practice D, "Nitric-Hydrofluoric Acid Test for Detecting Susceptibility to Intergranular Attack in Molybdenum-bearing Steels," ASTM A 262.[7] The mixed acids can differentiate sigma phase (which may cause low-carbon, molybdenum-bearing grades like 316L to *fail* a boiling 65% acid test, as in Practice B) from chromium carbide precipitation.

Phosphoric Acid-HF Mixtures

While pure phosphoric acid can usually be handled in Type 316L (S31603) stainless steel in conventional applications, fluoride contamination (as in wet-process acid) usually demands higher alloys, such as Alloy G-30.

References

1. N. Sridhar, Paper No. 19, "The Effect of Velocity on Corrosion in H_2SO_4-HF Mixtures," CORROSION 90, NACE, Houston TX.
2. V.S. Zotikov, Khem Prom Moscow 1975, p. 773.
3. Anonymous, *Materials Performance*, Vol. 30, No. 4, p. 55 (April 1991).
4. C.P. Dillon, "Quality Assurance Measures for Some Stainless Steel Products in Corrosive Service," *Materials Performance*, Vol. 30, No. 7, p. 54 (1991).
5. H.S. Jennings, CORROSION 2001, Paper 01345, NACE, Houston, TX.
6. *Metals Handbook*, 9th Edition, "Corrosion", p. 1155. ASM International, Metals Park, OH. (1987).
7. D. Warren, "Nitric-Hydrofluoric Acid Evaluation Test for Type 316L Stainless Steel," ASTM Bulletin, No. 230, May 1958, pp. 45–56.

7

Specific Equipment Requirements

Different types of process equipment may require specific attention to details relating to design, fabrication, and operation, as well as adequate corrosion resistance for the intended service. This chapter covers some details related to specific types of equipment as well as specific recommendations for such equipment, as determined by environment and service conditions. Suggested materials/environment combinations are summarized in Tables 7.1 through 7.5.

Tanks, Vessels, and Columns

Carbon Steel AHF storage tanks are pressure vessels, normally designed for 60 psig pressure and must be constructed in accordance with a code such as the ASME Boiler and Pressure Vessel Code, Section VIII, Div. 1 or 2. Steel may also be used for HF gas to 300°C (570°F). 70% HF (but not weaker concentrations) may be stored in steel tanks, with PWHT and a ¼-inch corrosion allowance, at ambient temperatures.

Steel vessels for all HF and AHF service should be constructed of clean steel (see further details below) and painted white to minimize solar heating.

In designing a tank, external attachments, including pads, must be vented to prevent the accumulation of hydrogen gas formed by corrosive attack on ferrous alloys. Post-weld heat treatment (PWHT) is strongly recommended although not required unless by the ASME Code, if the above requirements are met. However, consideration should be given to stress-relieving plates containing openings or supports before assembly and weld hardness must be controlled.

Consideration should be given to acoustic emission testing of large vessels, concurrent with hydrostatic testing, to provide baseline data for future inspections.

Steel Plate Specifications The recommended steel specification for stationary vessels is ASTM A516 with the following additional requirements:

- Plates shall be normalized.
- Sulfur 0.006% maximum.

Table 7.1 Summary for Materials of Construction for Anhydrous HF Service

Equipment	Material	Remarks
Tanks, Vessels, & Columns	Carbon Steel	Clean notch-tough steel, PWHT 615°C (1125°F) min.
	Alloy 400	Bubble caps, instruments. Thermal stress relief against SCC or use Alloy 600
Heat Exchangers	Carbon Steel	Passivated seamless steel tubes to 6 fps and 66°C (150°F) maximum.
	Alloy 400	Seamless tubes to 188°C (370°F). Avoid moist air to prevent SCC.
	70–30 Cu/Ni	Better erosion-corrosion resistance than Alloy 400. OK to 188°C (370°F)
Vaporizers	Alloys 400 or 600	Seamless tubes. Limit steam pressure to 50 psig to minimize corrosion.
Piping	Carbon Steel	66°C (150°F) and 6 fps max. Seamless A106, Gr A or B, to 4"; ERW A106, Gr A or B, >4". API 5L in larger sizes.
	Alloy 400	188°C (370°F) Max. Do not allow air ingress to drained lines.
Pumps & Compressors	Cast Steel	Ambient temp., intermittent service. Alloy 400 or Alloy 20 required.
	M-35–1 (400)	188°C (370°F) Max.
Valves	Cast Steel	Ambient; Alloy 400 trim
	CF8M, CF3M	Ambient, <0°C (<32F)
	M-35–1,CY-40	To 188°C (370°F)
	PTFE or PFA	As lining to 50°C (140°F) or as sleeve for steel plug cock with Alloy 400 plug.
	CN7M	To 100°C (212°F)
O-Rings	Fluorocarbon Elastomers	Specially cured and compounded. Ambient temp.
	Perfluorelastomers	Specially cured and compounded. To 100°C (212°F)
	PTFE	To 150°C (300°F)
Gaskets	PTFE	Sheet, w/o filler
		Sheet, CaF_2 filler
		Sheet, carbon filler
		Sheet, metal grid reinforcement
	Spiral-Wound	PTFE and flexible graphite fillers. Tape 304 or 316 (100°C) or high nickel alloys (200, 400, 600, or C-276)
Bolting	B-7	Controlled hardness
Shipping Containers	Steel	Notch-tough, HRC 22 max. normalized A516, Gr 70 or A537, Cl 1, PWHT 600°C (1100°F) min. Valves, etc. to NACE MR0175.

Table 7.2 Summary for Materials of Construction for HF Gas Streams at Elevated Temperatures

Equipment	Material	Remarks
Tanks, Vessels, & Columns	Carbon Steel	To 300°C (570°F). Avoid alloy steels.
	Type 304/316	To 150°C (300°F) Avoid grades with Cb, Ti.
	Alloy 600	To 500°C (930°F). Use ERNiCrMo4 rod although it age-hardens >400°C.
Heat Exchangers	Carbon Steel	To 300°C (570°F). Avoid alloy steels.
	Type 304/316	To 150°C (300°F) Avoid grades with Cb, Ti.
	70–30 Cu/Ni	To 300°C (570°F)
	Alloy 600f	To 500°C (930°F). Use ERNiCrMo4 rod although it age-hardens >400°C.
Piping	Carbon Steel	Seamless A106, Gr A or B, to 4"; ERW A106, Gr A or B, >4". API 5L in larger sizes. Avoid alloy steels.
	Type 304/316	To 150°C (300°F) Avoid grades with Cb, Ti.
	Alloy 600	To 500°C (930°F). Use ERNiCrMo4 rod although it age-hardens >400°C.
Compressors & Blowers	Carbon Steel	To 300°C (570°F). Use nickel alloy trim
	Alloy 600	To 600°C (1110°F). Use ERNiCrMo4 rod although it age-hardens >400°C.
	Alloy 242	To 600°C (1110°F)
Valves	Carbon Steel	To 150°C (300°F) with 18–8 trim; to 300°C (570°F) with Ni alloy.
	CF8M, CF3M	To 150°C (300°F)
	CY-40 (600)	To 500°C (930°F)
O-Rings & Gaskets	PTFE	CaF_2-filled, metal-grid or spiral-wound nickel-alloy gaskets
	OF Copper	To 500°C if no S.
	Alloy 201	To 500°C; w/o S.
Bolting	Carbon Steel	Per Code
	Alloy 718	For higher temperature

- Inclusion shape control (see Chapter 3, Section 2.1.2).
- Ultrasonic inspection ASTM A578, Level 2, S1 (modified to 1" diameter).
- S21 of ASTM A20, "Restricted Unspecified Elements."

HIC-tested A516 steel in grades 60, 65, and 70 is now offered by the large plate mills, with the following additional requirements:

- No micro-alloying elements (V,Ti,Cb, B) added, to avoid retained hardness after PWHT.

Table 7.3 Summary for Materials of Construction for 70% Hydrofluoric Acid

Equipment	Material	Remarks
Tanks, Vessels, & Columns	Carbon Steel	Ambient temperature. Corrosion allowance ¼″ Clean steel, PWHT, paint white for exposure to sunlight. (Same for CS with chlorobutyl lining.) 40–60°C (104–140°F). Vapor-phase attack above 32°C (90°F)
	Fluorinated Plastic-Lined Steel PFA Liner	Ring-supported; to 120°C (250°F)
	Brick Lined	Special Carbon Brick, PTFE membrane; to 110°C (230°F)
	Alloy 400	To 80°C (175°F), PWHT
Heat Exchangers	Alloy 400	To 80°C (175°F). Strength weld and stress relief vs. SCC.
	70–30 CuNi	To 80°C (175°F). More resistant to SCC.
	FEP, PFA	To 200°C (390°F) Limit pressure 20–40 psig shell, 40–90 psi on tube side
	Graphite	PTFE-impregnated, To 170°C (340°F) full vacuum; 230°C (445°F) wall temperature
Piping	PP-Lined Steel	Ambient temperature <40°C (108°C)
	Alloy 400	To 80°C (175°F), PWHT to avoid SCC. Exclude air.
	Fluoroplastic Lined Steel	PVDF 93°C (200°F) FEP 100°C (212°F) ECTFE 110°C (230°F) PTFE 150°C (300°F)
	CSPE, Butyl Rubber	Hoses, ambient temperature

(continued)

- Plate over 1.18″ (30 mm) thick tested for HIC susceptibly per NACE Standard TMO284 (modified).
- 0.002% max. sulfur.

Grade 55 is recommended for vessels not to be PWHT, but it is good practice to PWHT all vessels.

When Grades 65 or 70 are selected, the lowest carbon equivalent (CE) that will meet the minimum tensile strength (after stress relief) should be agreed upon with the steel mill.

$$CE = C + Mn/6 + (Cr+Mo+V)/5 + (Ni+Cu)/15$$

This will minimize hardening during welding and retained hardness after stress relief. Slight differences in steel mill practice, particularly the amount of scrap used

Table 7.3 (continued) Summary for Materials of Construction for 70% Hydrofluoric Acid

Equipment	Material	Remarks
Pumps	M-35–1	To 80°C (175°F)
	FP-Lined Steel or Ductile Iron	To 110°C or manufacturer's recommendation
Valves	M-35–1	To 80°C (175°F)
	FP-Lined Steel or Ductile Iron	To 110°C or manufacturer's recommendation
O-rings	Chlorobutyl CSPE	Ambient temperature to 50°C (122°F)
	Fluoroelastomers	To 70°C (158°F); proper curing
	Perfluoroelastomers	To 100°C (212°F); proper curing
	PTFE-Shielded Elastomer	To manufacturer's recommendations.
Gaskets	Chlorobutyl	Ambient temperature
	Spiral-Wound Alloy 400	PTFE or graphite filler; 80°C (175°F). Metal to metal only.
	Flexible Graphite PTFE	Unsintered, sheet or as envelope over elastomer.
Shipping	Carbon Steel	To usual special requirements
	Lined Steel	Chlorobutyl elastomer (painted white) or PVDF/PFA sheet-lined.

Table 7.4 Summary for Materials of Construction for Electronic Grade HF (49%)

Equipment	Material	Remarks
Tanks	PVDF or PFA	Solid, 4–10 gallons. Sheet-lined steel, 90-mil minimum. Backed with carbon or cloth and FRP.
Heat Exchangers	PFA	Tube-and-shell or coils within tank.
Piping	PVDF or PFA	Butt welded
Pumps & Valves	PVDF OR PFA	
O-Rings & Gaskets	Fluoroelastomer	Or rubber/elastomer with PVDF or PFA sheathing.
Drums	PFA	Rotomoled 55-gallon.
Tank Truck Trailers	PVDF or PFA	Fluoroplastic-lined carbon steel.

and its impurities, will affect the relationship between CE and tensile strength as will thickness. Consequently, it is necessary to reach agreement with the mill. This CE will normally be on the order of 0.43. The mill should be told that the PWHT will be at 1125–1200°F (600–650°C) for the time designated so that the mechanical properties will be so determined.

Table 7.5 Summary for Materials of Construction for Up to 50% Hydrofluoric Acid

Equipment	Material	Remarks
Tanks, Vessels, & Columns	Rubber-Lined	Soft to 32°C (90°F) max.; to 40%, semi-hard to 66°C (150°F).
	Fluoroplastic/FRP	To 60°C (140°F).
	Impervious Graphite	To 48% at B.P.; 60% to 85°C (185°F).
	Fluoroplastic-Lined Steel	To manufacturer's recommendations.
	Alloy 400	To defined limits. PWHT against SCC.
	C-Brick Lined	To 110°C (230°F) with PTFE membrane.
Heat Exchangers	Alloy 400	To defined limits. Avoid dissolved oxygen & PWHT against SCC.
	70–30 CuNi	Resists SCC, avoid dissolved oxygen.
	Impervious Graphite	Phenolic-impregnated to 48% at 160°C (320°F).
	Fluoroplastic	To 200°C (390°F) with prescribed pressure limitations.
Piping	Rubber-Lined	To prescribed limits.
	PP-Lined Steel	To 66–93°C(150–200°F)
	Alloy 400	See isocorrosion curve.
	Fluoroplastic Lined	To prescribed limits.
Pumps	PP-Lined Steel or Ductile Iron	To 60°C (140°F).
	Fluoroplastic-Lined	To prescribed limits.
	Impervious Graphite	48% to B.P.; 60% to 85°C (175°F)
	M-35–1	See isocorrosion curve
Valves	Fluoroplastic-Lined	To prescribed limits.
	M-35–1	See isocorrosion curve. Soft PTFE seats for globe shut-off; PTFE sleeves for plug cocks.
O-Rings	Chlorobutyl	66°C (150°F) to 50%; 90°C (195°F) $<$30% HF.
	CSPE	To 70°C (160°F)
	Fluoroelastomers	To 100°C (212°F) with proper curing agents.
	Perfluoroelastomers	To 110°C (230°F) with proper curing agents.
	PTFE-Shielded Elastomers	To prescribed limits.
Gaskets	Chlorobutyl	As above.
	Flexible Graphite	To 110°C (230°F).
	Spiral Wound	PTFE or graphite with Alloy 400 (metal-to-metal only).
	PTFE	Plain, filled (carbon or Ca_2), or unsintered.
Bolting	Steel	See specific requirements in text.

It should be noted that the addition of micro-constituents is not unusual, particularly in Europe, and that this can affect the post-stress relief hardness in the weld heat-affected zone (HAZ).

Steels with tensile strengths over 85,000 psi should not be used.

Fabrication and Testing Welding procedures should be selected to minimize the hardness of both welds and HAZ. "Wash passes" (frequently used to enhance radiography interpretation) should not be used nor should downhill welding be used on vertical seams. Submerged arc (SAW) and coated electrode (SMAW) welding are preferred. High-manganese flux must not be used in SAW welding, and low-hydrogen electrodes of $3/16''$ minimum diameter should be employed for SMAW welding.

Preheat is required as specified by code or by the welding engineer.

Weld procedures shall be qualified on plates of the same specification as to be used for the vessel. A hardness profile shall be made of weld and HAZs on a full cross-section metallurgical mount. Measurements shall be made on as-welded samples and on stress-relieved samples, if stress reliving is required. Maximum micro-hardness shall be the equivalent of Rockwell C 22 (HRC 22).

Following fabrication, micro-hardness checks of welds and HAZ (when required by the inspector) shall be made at intervals of 50 ft. for vessels fabricated of Grades 65 or 70. If harnesses of HRC 22 or higher are found, additional testing will be required to determine the extent of the problem and the appropriate corrective action to be taken.

Note that NACE Standard RP0472–87, which covers weld hardness control, recommends a maximum weld hardness of Rockwell B 200 (HBN 200) to prevent weld cracking of carbon steel welds in various hydrogen stress cracking environments, including hydrofluoric acid. However, in the case of AHF, weld and HAZ hardness below HRC 22 (HB 237) have been found to be acceptable. Experience has also shown that excessive weld hardness can be produced using SMAW coated electrodes (E7018), contrary to paragraph 2.2.1 of RP0472–87.

All butt welds shall be radiographed and shall meet the requirements of the applicable code, such as Section VIII of the ASME Boiler and Pressure Vessel Code.

All carbon steel vessels made of Grades 65 or 70 plates should have their internal welds initially inspected by wet fluorescent magnetic particle methods and again during 5 to 10 year scheduled internal inspections. Alternatively, consideration should be given to acoustic emission (AE) testing coincident with initial hydrostatic testing, in order to provide baseline data for future inspections. When necessary shear-wave ultrasonic inspection of the suspect locations should be conducted. An internal inspection should be made during the first five years and at least every ten years thereafter.

Stainless Steels Types 304 and 316 may be used *only* in dry gas streams to about 150°C (300°F). The Cb- and Ti-stabilized grades must not be used. Various types of corrosion will ensue in the event of moisture contamination.

The high-performance Alloys 20 and 825 are used only to cope with H_2SO_4/HF mixtures encountered in HF manufacturing. To avoid selective corrosion of Alloy 20 welds, the Cb-free 320LR filler metal is used for welding. Alloy 825, being stabilized with titanium, is less prone to selective weld attack.

Alloy 400 (N04400) The 70 Ni-30 Cu alloy is used in AHF to the critical temperature, especially for bubble caps in columns and for instruments. Thermal stress relief is advisable for welded, cold-formed, and cold bent parts to minimize the possibility of SCC in the event of ingress of moist air.

Alloy 400 is seldom used in dry HF at elevated temperature, although resistant to about 500°C (930°F), because of the higher strength of alternative alloys. The appropriate welding rod is low in Ti or Ti-free, such as RNiCu 5 or ENiCu2.

Alloy 400 is unsuitable for aqueous HF solutions below 70% except in the total absence of air or oxidants (which can rarely be achieved in practice). When used, it should receive PWHT at 675–730°C (1250–1350°F) to minimize the possibility of vapor phase SCC in the event of ingress of air.

Cupronickels are preferred in many instances because they do not suffer SCC by cupric fluoride under shutdown or upset conditions.

Alloy 600 (N06600) Alloy 600 is more expensive than Alloy 400 but is suitable in AHF up to the critical temperature and is more resistant to possible SCC. It is the preferred alloy for high-temperature HF-bearing gas streams to 500°C (930°F). The weld filler metals contain small amounts of Cb or Cb-Ti and may suffer preferential attack. The Ni-Cr-Mo Alloy C-276 rod (ER NiCrMo 4) has been successfully used, although it age-hardens above 400°C (750°F).

Alloy 600 is not used for aqueous solutions.

Alloys 200 (N02200) and 201 (N20100) These materials are not used for AHF, being more expensive and more difficult to weld than Alloy 400. Likewise, although the Alloy 201 is metallurgically stable above 300°C (570°F) and resists AHF streams to 500°C (930°F), it is generally unsuitable. This is due to its poor strength and to *severe* corrosion due to sulfur contamination in many high-temperature HF environments. Also, the weld rod (ERNi3) contains 3% Ti which is about 800–1000 times less resistant than the parent metal.

Nonmetallic Materials Lined steel structures are commonly used for aqueous HF service. The various plastics, elastomers, and carbon brick are selected to cope with specific operating temperatures and have been described in Chapter 4. The plastic materials are also used as freestanding vessels, when appropriate.

Lined or Coated Steel Vessels The successful use of these materials as linings for immersion service depends upon design and fabrication. The surface finish of the vessel is critical, as well as the material selected, its thickness, application to the surfaces of the vessel, and testing for imperfections or holidays.

The vessel must be designed with sufficient thickness for the pressure and liquid head and necessary stiffness and, if lined with brick, for the weight of the lining. The steel should have sufficient notch-toughness to prevent brittle fracture if stresses should build up as a result of corrosion of the shell. Selection of the specific grade of steel will depend upon the lowest expected temperature. For nonrefrigerated outdoor tanks, selection can be based on either the ASME Code for Unfired Pressure Vessels, or, if not a pressure vessel, on API Standard 620.

Design and fabrication details are covered in NACE Standard RP0178–89 Standard Recommended Practice, "Fabrication Details, Surface Finish Requirements and Proper Design Considerations for Tanks and Vessels to be Lined for Immersion Service." Some of the more important points are as follows:

- The welds must be either ground smooth and flush, as in the case of many sheet linings (NACE Weld Preparation Designation A or B), or the welds ground smooth (but not flush) and blended into the plate surface (NACE Weld Preparation Designation C), as required for fused linings.
- Outlets must be flanged or pad type, rather than threaded.
- Outlets must be flush with the inside of the vessel and attached with full-penetration welds. Sharp edges on the process side must be ground to a minimum radius of 1" for sheet linings and 4" for fused linings.
- Butt welding should be used instead of lap welding. However, where lap welds must be used (as in the bottom and/or roof of large field-erected storage tanks), double (inside and out) continuous fillet welds, ground smooth on the inside, may be used.
- All construction involving pockets or crevices, or that cannot be properly abrasive blasted and lined, must be avoided.
- Stiffening members should be on the outside of the vessel or tank.
- The openings must be of large enough diameter/pipe length to allow adequate surface preparation and lining.
- There must be a large enough manhole (18" diameter minimum) to provide access.
- The internal surfaces must be given a white metal sand- or grit-blasting and an initial coating of adhesive or prime coating immediately.

Elastomer linings are usually 3⁄16 inch thick, but ¼ inch thick linings will have a longer life. Sheets should be made with more than 3 calendared plies. Lap joints should be between ½ inch and 1¼ inch in length. Spark testing should be carried out, the test voltage being determined by the characteristics of the elastomer. In the case of natural rubber linings, 20,000 volts is usually used.

Plastic linings should be thick enough to resist permeation. No lining less than 40 mils can be considered as an immersion corrosion barrier. In the case of sheet linings with glass or polyester cloth backing (e.g., high-purity PVDF or PFA), the total thickness should be a minimum of 90 and 60 mils, respectively. These linings are normally tested at 10,000 volts at the fabricators. Subsequent field testing should be done at lower voltages or with a low-voltage holiday tester, since high-voltage testing can cause defects and must not be done repeatedly or unnecessarily.

Plastics and Elastomers Below 70% HF, electronic grade 49% acid may be handled in pigment-free, high-purity PFA and PVDF. These are employed because of increased demand for extreme purity in processing 1M bits and 4M bits using submicron technology. These fluoroplastics are replacing the HDPE previously used.

Glass or carbon cloth-backed PFA or PVDF, externally armored with FRP, are more economical than lined steel construction, although at some sacrifice in safety. Code design requirements must be met and a conductive layer is needed to permit spark testing.

PTFE non-bonded liners in steel shells have been used for 70% HF to 120°C (250°F), supported by internal graphite rings to resist diametric shrinkage (e.g., 36" ID and above). Lacking internal support, there is a potential danger of collapse under vacuum.

For other HF solutions up to 50%, similar materials are useful. In addition to the fluorocarbons, lesser plastics may be used, as follows:

- HDPE may be employed at ambient temperatures, using a cross-linked polymer. Nozzle stresses must be minimized and no parasitic stresses imposed. Safeguard against potential releases.
- Polypropylene armored with FRP is acceptable to 60°C (140°F) if potential safety problems are safeguarded. It is essential to meet Code design requirements, particularly in regard to nozzles and flanges. A conductive layer must be incorporated to permit spark testing.
- UPVC, reinforced with FRP, is limited to a maximum of 40% concentration and 50°C (122°F). Code and safety requirements must be met. A conductive layer must be incorporated to permit spark testing.
- Vinyl ester FRP, with a carbon veil inner layer, may be used to 30% HF and 60°C (140°F) with the same precautions as above.
- Carbon-filled phenolic, armored with FRP, may be used to 110°C (230°F) with due consideration of the safety aspects of an inherently brittle vessel.

For up to 50% acid other than the Electronic Grade, elastomeric linings in steel may be used within the following limitations:

- Soft natural rubber −32°C (90°F) max. (limited life)
- Semi-hard natural rubber −66°C (150°F) max. to 40% HF
- Chlorobutyl elastomer −66°C (150°F) max.; 90°C (195°F) to 30% HF

Carbon and Impervious Graphite In 70% acid, graphite rings have been used in PTFE-lined columns as already described. In this service also, carbon brick linings over a PTFE membrane are used to 110°C (230°F). A silica-free, carbon-filled phenolic mortar is used to cement the low-ash brick.

Phenolic-impregnated impervious graphite columns have been used to the atmospheric boiling point for 48% and to 85°C (185°F) for 60% HF. This type of brittle construction requires safeguarding by isolation or other means.

Heat Exchangers

Heat exchangers of various kinds are utilized in HF services (i.e., vaporizers, heaters, coolers, and condensers) and are fabricated from different materials of construction, as required by specific process conditions. The materials selected must, of course, be compatible with the utilities medium (e.g., steam or water) on the side opposite the process.

All tubing and other components must be fabricated to the appropriate codes, such as TEMA and the ASME Code for Unfired Pressure Vessels.

Carbon Steel Carbon steel exchangers should be assembled with seamless tubes. They are used to a maximum temperature of 66°C (150°F), with a limiting velocity of 6 fps, in AHF. Passivation, as described in Chapter 3, should be specified to ensure an adequate protective film following any water washing.

At elevated temperatures, steel may be used in HF gas to about 300°C (570°F). HIC is unlikely particularly above 120°C (250°F), but chromium-bearing alloy steels should be avoided.

Stainless Steels Types 304 and 316 stainless steels may be used in *dry* HF only. One must avoid grades containing Nb, such as Types 347 and 309Nb or Ti such as Type 321.

Alloy 400 (N04400) Alloy 400 seamless tubes may be used to the critical temperature* of 188°C (370°F). Washing and drying (or other protection from exposure to moist air during shutdowns) is required to prevent general corrosion and SCC, as discussed in Chapter 3.

Alloy 400 is seldom used for hot HF-containing gases (although its corrosion resistance is comparable to Alloy 600) because of lower high-temperature strength and possible SCC. Weld filler metals must be free of or low in Ti.

Regarding wet HF, Alloy 400 may suffer SCC if any vapor phase is exposed to moist air. If necessary, one may consider stress-relieved, strength-welded exchangers.

Cupronickel (C71500) 70–30 cupronickel (C71500) exchangers may also be used to the critical temperature. Their corrosion and erosion-corrosion resistance is second only to Alloy 400 (N04400). They have an advantage over the 70 Ni–30 Cu alloy in their high degree of resistance to SCC.

Alloy 600 (N06600) Alloy 600 tubes may likewise be used to 188°C (370°F). Although more expensive than Alloys 400 and 70–30 cupronickel, they are immune to SCC.

---

*NOTE: The critical temperature is the highest temperature at which AHF can exist as a liquid, i.e., 188°C (370°F). At this temperature, the corresponding pressure is 940 psig. Other temperature/pressure relationships are 60°C (140°F)/58 psig and 100°C (212°F)/150 psig.

Vaporizers are subject to some specific limitations. Alloy 600 (N06600) seamless tubes are used, with the steam pressure limited to 50 psig to minimize corrosion by deposits of salts and sulfuric acid left after evaporation of liquid HF. The Ni-Cr-Fe alloy is marginally more resistant than Alloy 400. Alloy 400 (N04400) seamless tubes are less expensive than Alloy 600 (N06600) and are subject to the same steam temperature limitation.

Alloy 600 is the most widely used alloy for high temperature HF environments. Filler metals normally used contain Cb or Cb + Ti, which can cause preferential corrosion.

Alloy 800 (N08825) Alloy 825 seamless tubes have a limiting temperature of 100°C (212°F). This alloy has excellent resistance to mixtures of HF and concentrated sulfuric acid to 80°C (175°F) and is widely used throughout HF manufacturing plants.

Nonmetallic Heat Exchangers For aqueous hydrofluoric acid solutions, heat exchangers of plastic, carbon, or ceramic may be employed.

FEP or PFA exchangers to the relevant code may be used to a maximum wall temperature of 200°C (390°F). Pressure limitations are 20–40 psig on the shell side and 40–90 psig on the tube side. PFA tubes may be carbon-filled to improve heat transfer.

PTFE-impregnated graphite exchangers, fabricated to the relevant ASME code are rated to 100–150 psig and to full vacuum at 170°C (340°F) for block type units totally enclosed in a steel shell. The limiting wall temperature is 230°C (445°F). Because this material may not be completely impermeable, the application should be discussed with the manufacturers.

For hydrofluoric acid, phenolic-impregnated graphite may be employed up to the atmospheric boiling point in 48% acid. The maximum wall temperature is 160°C (320°F). In 60% acid, the limiting temperature is 85°C (185°F).

Piping

Carbon Steel Steel piping is normally used for liquid AHF service to about 66°C (150°F) and for gaseous AHF to about 300°C (570°F). Seamless ASTM A106, Grade A or B is preferred; however, welded carbon steel grades have also been used successfully. For liquid flow at ambient temperature, the limiting velocity is 6 fps (2 m/s).

- For sizes through 2″ ips S/80 is required. For sizes up to and including 4 inch, electric-resistance welded pipe in accordance with ASTM A587 has been used, although A106 Grade A or B seamless is preferred.
- For large size pipe (6″ and over), it may be permissible to use electric resistance-welded, e.g., ASTM A53 (Grade A or B) or API-5L. The piping should be made from plate meeting the plate specifications described in 1.1 above or NACE Standard TMO284, "Test Method for Evaluation of Pipeline Steels for Resistance to Stepwise Cracking."

The piping must be butt welded rather than threaded. Weld-neck flanges are preferred. Rolled flanges and slip-on flanges should be avoided, because hydrogen can accumulate between the welds of the latter. Normalized flanges should be required.

Welding procedures must be qualified for each grade of piping used in accordance with the ASME Code, Section IX.

Weld or HAZ hydrogen stress cracking (HSC) has not been reported in piping butt welds. In external multipass welding, the hardness of the root pass and adjacent HAZ is lowered by the annealing effect of subsequent passes. Also, the interior weld contains residual compressive stress from shrinkage of the cover passes. To avoid hydrogen-induced cracking (HIC) during welding, use low-hydrogen electrodes and preheat to 38°C (100°F) if the metal temperature is less than 10°C (50°F).

A recent study by The Welding Institute (TWI) (Cambridge, UK) has confirmed that the hardness limits for external pipe welds required by NACE Standard MR0175 ("Sulfide Stress Corrosion Cracking Resistant Metallic Materials for Oilfield Equipment") can be relaxed from HRC-22 to HRC-30. Since HF is a somewhat less severe environment than H_2S, a hardness requirement for external piping welds is probably unnecessary, if all the other requirements discussed above are followed.

PWHT of carbon steel piping is seldom required. Occasionally, fittings are received with a hardness greater than HRC 22 and heat treatment may then be considered an option.

For all liquid AHF services, and gaseous HF above 5 psig, 100% radiography is required, per ANSI B31.3. Fittings must provide a butt weld configuration.

For all low-pressure AHF gas (5 psig or less), a minimum of spot radiography in accordance with ANSI B31.3 is required.

Steels and low-alloy steels are nonresistant below 70% acid strength.

Stainless Steels Only the high-performance alloys, such as Alloy 20 or Alloy 825, as seamless pipe made to meet the requirements of ANSI B31.3 are suggested for AHF. Generally, the Alloy 825 is more available as seamless pipe. The temperature limit is 100°C (212°F).

For HF gas at elevated temperatures, conventional stainless may be employed, made to meet the ANSI requirements. Types 304 and 316 may be employed but not the niobium- or titanium-bearing grades such as Type 347, 309Nb or 321.

Alloy 400 (N04400) The 70 Ni–30 Cu alloy may be used to the critical temperature in AHF, but ingress of air (as to drained lines) must be prevented. It is not normally used in HF gas at elevated temperature.

Alloy 400 may be used in aqueous solutions only if air or other oxidizing species are rigorously excluded. Purging must be effected with dry nitrogen, and cold-worked bends should be PWHT at 675–730°C (1250–1350°F) to minimize the possibility of SCC.

Alloy 600 (N06600) Alloy 600 (N06600) is widely used for high-temperature HF applications because it has higher strength than the 70 Ni–30 Cu alloy. It should be noted that weld filler metals containing Cb or Cb + Ti can suffer preferential corrosion. Alloy 600 (N06600) is not to be considered for aqueous hydrofluoric acid.

Nonmetallic Piping Systems Polypropylene-lined steel pipe may be used for up to 70% hydrofluoric acid at 42°C (108°F). For larger sizes than are available in the PP-lined steel, chlorobutyl rubber-lined pipe may be used to 60°C (140°F) in 31–50% HF and to 90°C (195°F) up to 30% concentration.

Fluoroplastic-lined steel piping systems may be employed in aqueous hydrofluoric acid, as follows:

- PVDF to 93°C (200°F)
- FEP to 100°C (212°F) up to 130°C (265°F) with thick liner, but limited vacuum capability
- ECTFE to 110°C (230°F)
- PTFE to 150°C (300°F); consider limiting to 110°C (230°F) and 50 psig because of HF hazards.

Chlorosulfonated polyethylene (CDPE) or butyl elastomer *hoses*, with a tough elastomer cover, are used for loading and unloading aqueous hydrofluoric acid (e.g., for tank cars and tank trailers).

Pumps and Compressors

Compressors and blowers are used to convey gas or vapor streams while pumps move liquid environments.

Steel and Cast Iron Carbon steel, but not alloy steels, may be used for HF vapors to 300°C (570°F) provided nickel alloy trim is employed for greater erosion-corrosion resistance. Steel pumps are not suitable for liquid HF, although a cast steel with alloy trim (N04400 or N08020) will give a limited service life in AHF at ambient temperature. These are sometimes used for loading/unloading or to feed AHF to a batch process.

Stainless Steels Although not resistant to aqueous HF, CF8M (J92900) and CF3M (J92800) pumps have been used for AHF at ambient temperatures or below, with a limited service life. They have been used at ambient temperature in a manner similar to steel pumps, as discussed above, and for continuous service in refrigerated AHF.

The cast Alloy 20, CN7M (N08007), has been used in AHF only to as high as 100°C (212°F). This material is difficult to cast and should be bought only from a prequalified supplier. At a very minimum, the casting should meet MSS Visual Quality Standard SP-55. Preferably, dye penetrant and/or radiographic inspection per ASTM E-94, E-142 or ASME Code Section V should be required, the acceptance limits for "discontinuities" being negotiated with the supplier.

Alloy 400 Alloy 400 and Alloy 600 are the primary materials recommended for HF service under the velocity conditions that occur in pumps and similar movers.

In the cast form as M-35–1 (N04020) this material may be used for AHF to the critical temperature. It is the alloy of choice for HF alkylation processes wherein some water may be formed. Obtained from a prequalified foundry, it is also employed for up to 70% HF (MS Visual Quality Standard SP-55). The temperature limits are defined by the isocorrosion curve in Figure 4.3.

Alloy 600 In the cast form as CY-40 (N06040), the Ni-Cr-Fe alloy has been used infrequently to the critical temperature in AHF when there has been concern over SCC of the M-35–1. This is the preferred alloy for hot HF gas streams to 600°C (1110°F), being less expensive and more thermally stable than higher nickel-based alloys described in previous chapters. It is not used for aqueous acid.

Nonmetallic Materials For up to 70% HF, steel or ductile iron pumps lined with fluorinated plastics such as PVDF, ECTFE, ETFE, PTFE, or PFA are useful to about 110°C (230°F). They are available in a range of sizes, and individual manufacturers may indicate other temperature limitations. These pumps employ mechanical seals (single or double type) with no metals in contact with the pumped liquid. A range of seal faces and rotating seat materials are available, including carbon, carbon-filled PTFE, and pure alumina (Al_2O_3). Cobalt-bonded carbides are not suitable.

Pumps for the 49% Electronic Grade are limited to PVDF and PFA.

For other HF concentrations up to 50%, polypropylene-lined steel or ductile iron pumps may be used to 60°C (140°F).

Valves

Valves include several designs and may be used either for throttling or shutting off the liquid flow. A number of different materials may be employed, depending on the process conditions.

Ferrous Alloys In HF service, steel and iron are only used with alloy trim or fluoroplastic components, as described further below.

The 18–8 grades of stainless steel, such as CF8M (J92900) and CF3M (J92800), may be used for AHF at and below ambient temperatures and in hot HF vapors to 150°C (300°F). They have no application in aqueous HF.

The Alloy 20 cast, CN7M (N08007), may be used to 100°C (212°F) in AHF with the same restrictions described for cast valves of this material (see above). However, the wrought Alloy 20 is sometimes employed as trim in cast carbon steel valves for ambient temperature service.

Alloy 400 (N04400) Alloy 400 and Alloy 600 are the preferred materials for valves in many HF services. The wrought Alloy 400 is used for trim in cast carbon steel valves for AHF at ambient temperature, while the cast M-35–1 (N04020) can be used up to the critical temperature. The latter is also used in aqueous HF (except for Electronic Grade) up to 70% to 80°C (175°F). Such valves are available in all types of valves and plug cocks, the latter with PTFE sleeves. Globe valves with a soft PTFE seat are recommended for shut-off service. As with pumps, the temperature limits are defined by the lower isocorrosion curve shown in Figure 4.3. Sometimes Alloy B-2 bellows seals are used when there is concern over the SCC propensity of Alloy 400.

Alloy 600 Cast Alloy CY-40 (N06040) valves may be used in AHF to the critical temperature. They are also used in lieu of M-35–1 (N04020) in hot HF vapors to 500°C (930°F). The Ni-Cr-Fe alloy finds no application in aqueous HF.

Nonmetallic Valves Both solid plastics and plastic liners can be employed in HF valving. For AHF, a cast carbon steel plug cock with a PTFE sleeves and an alloy plug (e.g., Alloy 400 or Alloy 20) may be used at ambient temperature. Carbon steel or ductile iron valves lined with PTFE or PFA may be used to 60°C (140°F), although there may be problems with permeation above that temperature's corresponding pressure.

For aqueous HF up to 70%, the fluorocarbon-lined carbon steel or ductile iron valves may be used to about 110°C. However, the 49% Electronic Grade requires PVDF or PVA of very high purity, as previously described.

O-Rings and Gaskets

Suitable materials for gaskets and O-rings are determined by the parameters of concentration and temperature of the HF environment. The appropriate materials are prescribed for the several grades of HF as follows:

Anhydrous Hydrogen Fluoride (AHF)

O-Rings:

- Fluorocarbon elastomer—ambient temperatures (Compounding and curing agent must be those specifically recommended by the manufacturer for AHF.)
- Perfluoroelastomer—100°C (212°F) (Compounding and curing as above.)
- PTFE—150°C (300°F)

Gaskets

- PTFE sheet (without filler)
- PTFE sheet with calcium fluoride or carbon filler
- PTFE sheet with metal grid reinforcement
- Spiral-wound gaskets with PTFE and flexible graphite fillers. The tape materials may be Types 304 or 316 to 100°C (212°F) max., nickel alloys (200, 400, 600) or Alloy C-276.

Hot HF Gas Streams Only metallic alloys are suitable above the temperature limits for fluorinated plastics and elastomers.

O-Rings

- Nickel 201 or copper (oxygen-free; C12000) to 500°C (930°F). For *sulfur-free* atmospheres only. Deoxidized copper may also be used.

Gaskets

- (PTFE to 200°C or 390°F)
- PTFE sheet with calcium fluoride filler
- PTFE sheet with metal grid reinforcement
- Spiral-wound gaskets with PTFE or flexible graphite fillers. The tape material should be a nickel alloy (400,600, or C-276).

70% Hydrofluoric Acid

O-Rings

- Fluoroelastomers—70°C (158°F) (with correct curing agents and additives)
- Perfluoroelastomers—100°C (212°F) (with correct curing agents and additives) PTFE shielded elastomer
- CSPE—50°C (122°F)
- Chlorobutyl—at ambient temperature

Gaskets

- Chlorobutyl—at ambient temperature
- PTFE sheet (plain or filled with carbon or calcium fluoride)
- Flexible graphite—110°C (230°F)
- PTFE envelope type with elastomer core to 110°C (230°F)
- Unsintered PTFE
- Spiral-wound gaskets with PTFE or flexible graphite filler and Alloy 400 tape—80°C (175°F); for metal-metal only.

Aqueous HF to 50%

O-Rings

- Fluoroelastomers—100°C (212°F) for 1–48% with correct curing agents and additives.
- Perfluoroelastomers—110°C (230°F) with correct curing agents and additives.
- PTFE shielded elastomer
- CSPE (chlorosulfonated polyethylene elastomer)—70°C (159°F)
- Chlorobutyl elastomer—66°C (150°F) for 40–50% HT; 90°C (194°F) for up to 30 HF.

Gaskets

- Chlorobutyl elastomer—66°C (150°F) for 40–50% HF; 90°C (194°F) 0 to 30% HT.
- PTFE sheet (plain or filled with carbon or calcium fluoride)
- Flexible graphite—110°C (230°F).
- Unsintered PTFE
- Spiral-wound gaskets with PTFE or flexible graphite filler and Alloy 400 tape. For temperature ranges, refer to see Figure 4.3, lower curve (for metal-metal only).

Bolting

Bolting for pipe and vessel closures, as well as for purchased valves, pumps, and instruments, is critical in view of the vulnerability of hardened carbon and alloy steels (as well as some stainless steels) to hydrogen charging and resulting hydrogen embrittlement (HE) from corrosion by hydrofluoric acid (see Chapter 4).

It is important that bolting be properly specified when ordered and subsequently checked upon receipt. Also, only acceptable bolting should be stocked. Erroneous marking of bolting is not uncommon and random hardness testing is recommended.

Although bolting is not generally directly exposed, small leaks are likely to result in failure from HIC if susceptible bolting is used. This is a particular problem with valve bonnets, since manufacturers often use hardened alloy steel studs (typically ASTM A193 Grade B7).

There have been cases in which the bonnet of a valve in AHF service has been lifted off as a result of HIC of A193 Grade B7 studs. Allen head set screws, made of hardened AISI E52100 alloy steel, have broken in pumps.

Although Grade B7M is recommended and should be required for valves, pumps, instruments, and internals, this grade is often not used where hard gaskets such as spiral-wound are encountered because of the likelihood of over-torquing during flange make-up. The most widely used bolting for these conditions is Grade B7, despite its susceptibility to HIC.

One European-based company imposes a hardness limit of 300 Vickers (HRC-30) and has had no reported failures.

When using Grade B7 bolting the following restrictions should apply:

- Although there is no maximum hardness specified in ASTM A193, it may be possible to negotiate one if bolting is purchased in quantity. In any case, it should be possible to sample incoming stock and control its use. The tensile strength of B7 bolts of, for example, 1" diameter, quenched and tempered at 590°C (1100°F), should be about 150 ksi, which corresponds to a hardness of HRC 33.
- Leak test the joint before introduction of HF.
- Observe for leaks daily or more often. Change out all bolts on a leaking flange and eliminate the leak.

Strain-hardened stainless steel bolting has also been used in AHF service and is permitted for use in tank cars. However, such bolting should not be used in chloride environments. Chloride SCC may result. The preferred material is ASTM A193 Grade B8M, Class 2, which is Type 316 (531600) stainless steel, strain-hardened to a maximum hardness of HB 320 (HRC 35). Grade B8, Class 2 (Type 304) has been used but is less stable metallurgically and forms alpha martensite upon strain-hardening, the amount depending upon the nickel-to-chromium ratio and the amount of strain-hardening. The presence of excessive amounts of martensite results in susceptibility to hydrogen embrittlement (HE).

Most nickel alloys can be used for bolting. However, Alloy 400, Alloy K500, and modifications of these alloys may suffer SCC and should be avoided. High-strength precipitation-hardened Alloy 718 has also been used.

Preferred steel and alloy steel bolting specifications are as follows:

Low Strength Bolting (Suitable for PTFE Sheet Gaskets)

Bolt—ASTM A-307, Grade B, Hex. Head
Nut—ASTM A563, Grade A, Heavy Hex. Head

Intermediate Strength—Normal Temperature Service

Pumps, Valves, Instruments, and Internal Connections

- Stud—ASTM A193, Grade B7M (100% hardness testing-HBN 99 (HRC22) max.)
- Nut—ASTM A194, Grade 2HM

Piping and Vessel External Flange Closures

- Stud—ASTM A193, Grade B7M (100% hardness testing-HBN 99 (HRC 22) max.) Preferred if over torquing is avoided.
- ASTM A193, Grade B7 (monitor hardness and leaks)
- Nut—ASTM A194, Grade 2HM

Intermediate Strength Low Temperature Service; −73°C (−99°F)

Pumps, Valves, Instruments, and Internal Connections

- Stud—ASTM A320, Grade L7M (100% hardness testing, HBN 99 (HRC 22) max
- Nut—ASTM A194, Grade 2HM (impact test per ASTM A320)

Piping and Vessel External Flange Closures

- Stud—ASTM A320, Grade L7M (100% hardness testing, HBN 99 (HRC22) max.
- ASTM A320, Grade L7 (monitor hardness and leaks)
- Nut—ASTM A194, Grade 2HM (impact test per ASTM A320)

Bolting Specifications that **should not** be used for HF piping, valves, pumps, vessels, and instruments include the following:

- ASTM A193 Grades B6, B6X, and B16
- ASTM A354
- ASTM A437
- AISI E52100
- SAE J429 Grades 5, 7, 8, 8.1, and 8.2

Tank Cars and Tank Motor Vehicles

The materials and fabrication requirements for tank cars and tank motor vehicles are covered by Department of Transportation "Hazardous Materials Regulations Pertaining to Hydrofluoric Acid and Hydrogen Fluoride" (49CFR). Unlined tank motor vehicles are covered in Part 178 and unlined tank cars in Part 179. The requirements for both parts are essentially identical:

- Plates for the tank shell, heads, and manway must comply with ASTM A516, Grade 70 normalized or ASTM A537, Class 1. Because blistering has been quite common, the HIC-resistant quality is recommended.
- All tank welds must have full penetration and all tank butt welds must be fully radiographed.
- Tanks must be PWHT at 595°C (1100°F) minimum. A range of 615–650°C (1125–1200°F) is recommended.
- The maximum hardness of the weld and HAZ after PWHT, measured on a production test plate cross-section, shall not exceed HRC 22.

Anhydrous Hydrogen Fluoride Only carbon steel is used for railroad tank cars and over-the-road tank trucks for AHF. The applicable codes, e.g., DOT-114A400W for the tank cars and MC 311 or MC 312 for the vehicles, must be invoked. Applicable requirements for the plate and the fabricated vessel were described above.

Valves, valve parts, fittings, and other accessories must meet the requirements of NACE Standard MR0175 (current edition) and be approved for HF service. No martensitic or ferritic stainless steels should be utilized. Bolting requirements are as described above.

70% Hydrofluoric Acid Both bare steel and lined construction are used for the transport of 70% HF in railroad tank cars and in over-the-road tank trucks, with specific precautions.

Steel For carbon steel mobile vessels, the steel must meet DOT specifications (114A100 F-2, W-2 and W-4 for tank cars; M-310, M-311 and MC 312 for tank motor vehicles). These requirements include the use of notch-tough steels (A-516 Gr.70 and A-537, Gr.1) suitable for low ambient temperatures and PWHT at 600°C (1100°F) min. As noted above, the HIC-resistant grade should be employed and hardness control of the weld HAZ is required. Valves, valve parts, and fittings must be in accord with NACE Standard MR0175 (latest edition), suitable for HF service. There are special bolting requirements, as described above. These containers must have a ¼″ corrosion allowance and be passivated. Even with these provisions fulfilled a corrosion rate of about 40 mpy must be anticipated.

Chlorobutyl Elastomer-Lined Steel Tank cars and trucks may be lined with chlorobutyl elastomer, in accordance with DOT specifications. They must be painted white to minimize the temperature rise occasioned by solar heating. This step can reduce the wall temperature by as much as 32°C (58°F) and prolong the life expectancy considerably. Such liners fail in the vapor space, the concentration of the vapor commonly exceeding 90%.

PVDF- and PFA-Lined Containers Glass-backed PVDF and PFA sheet-lined carbon steel construction must be fabricated to the relevant DOT specification. Both are being used to line tank truck trailers for HF service. The advantage of this type of construction is the smooth, easily cleaned surface and elimination of contamination problems.

Electronic Grade 49% HF This specialty product, requiring the highest possible purity, is shipped only in drums (Rotomolded PTFE is being replaced with high-purity PFA) and in the PVDF- or PFA-lined vessels. At one time, soft natural rubber linings were employed but they were unable to meet the increasing demand for higher product quality.

Dilute Hydrofluoric Acid For other aqueous solutions up to 50%, tank cars and trucks can be lined with elastomeric linings to the relevant code, such as DOT. Although a soft natural rubber lining in a steel vessel (painted white) has been used, a limited life is obtained. The preferred elastomeric lining today is the chlorobutyl variety.

The glass-backed PVDF or PFA sheet-lining approach described above is used for improved reliability in the higher concentrations.

Index

The Materials Technology Institute of the Chemical Process Industries, Inc. (MTI), founded in 1977, is a unique, not-for-profit technology development organization representing private industry. It sponsors projects focused on both developing new technology and transferring existing knowledge to day-to-day practice. Practical, generic, nonproprietary studies are conducted on the selection, design, fabrication, testing, inspection, and performance of materials of construction used in the process industries. The scope of work includes evaluation of metallic, and non-metallic materials, optimum design applications, fitness-for-service, mechanical integrity and life cycle determinations, and economic factors affecting performance of vessels, tanks, piping and other components.

Through membership and networking within MTI, companies can *access solutions* to nonproprietary problems of major concern to the process industries. Members can *capitalize* on the extensive expertise of member company representatives, *leverage* their technology investment by participating in the direction and results of MTI projects, and *utilize* MTI's books, reports, software and video training programs immediately as needed. Benefits to member companies are increased plant integrity, reliability and profitability.

James M. Macki, Executive Director
Materials Technology Institute, Inc.
1215 Fern Ridge Parkway, Suite 206
St. Louis, MO 63141-4405 USA
Tel: 314.576.7712
Fax: 314.576.6078
Web site: www.mti-link.org
E-mail: mtiadmin@mti-link.org